3/8/12
$136.00

Environmental Urban Noise

Advances in Ecological Sciences

Objectives

The objective of this series is to provide up to date information on basic and applied research and practical applications of a wide range of topics related to Ecology.

The series consists of books concerned with the state of the art information on ecological problems and as such comprises several volumes every year covering the latest developments and applications. Each volume is composed of authored works or edited chapters written by leading researchers and experts in the field.

The aim is to encourage and facilitate the interdisciplinary communication amongst scientists, engineers, economists and professionals working in the different areas of ecological research and applications. The scope of the series covers almost the entire spectrum of ecological sciences including ecological modelling, socioeconomic ecology, conservation, management and recovery of endangered and degraded areas, sustainable development, information techniques for development, ecological engineering, health and development, urbanism, and land use and legal and ethical problems.

Series Editors

J-L. Usó
Universitat Jaume I
Campus Peyneta Roja
Department of Mathematics
12071 Castellon
Spain

B.C. Patten
University of Georgia
Institute of Ecology
7112 Biosciences Building
Athens, Georgia 30602-2602
USA

C.A. Brebbia
Wessex Institute of Technology
Ashurst Lodge, Ashurst
Southampton, SO40 7AA
UK

Environmental Urban Noise

Edited By:

Amando García
University of Valencia, Spain

WITPRESS Southampton, Boston

Edited by:
Armando Garcia
University of Valencia, Spain

Published by

WIT Press
Ashurst Lodge, Ashurst, Southampton, SO40 7AA, UK
Tel: 44 (0) 238 029 3223; Fax: 44 (0) 238 029 2853
E-Mail: witpress@witpress.com
http://www.witpress.com

For USA, Canada and Mexico

WIT Press
25 Bridge Street, Billerica, MA 01821, USA
Tel: 978 667 5841; Fax: 978 667 7582
E-Mail: infousa@witpress.com
http://www.witpress.com

British Library Cataloguing-in-Publication Data

A Catalogue record for this book is available
from the British Library

ISBN: 1-85312-752-3
ISSN: 1369-8273

Library of Congress Catalog Card Number: 00-107712

*The texts of the papers in this volume were set
individually by the authors or under their supervision.*

Contents

Chapter 5 Prediction of urban noise
M. Arana

Chapter 6 Urban noise control
A. García

Preface

Environmental noise has become one of the greatest sources of nuisance in developed countries. This type of noise, briefly defined as unwanted sound, fills everything and affects everybody. People are constantly exposed to varying noise levels in their everyday lives, for instance, when working, using transport, resting at home or during leisure activities.

Noise pollution is closely related to human activity and generally occurs where this activity is concentrated, typically in urban areas (occupational noise, a most important problem which affects millions of workers around the world, will be not considered here). The extent of this problem is large. For example, about 25% of the European population is exposed, in one way or another, to daytime transportation noise levels exceeding 65 dBA. Sound levels as high as 70-80 dBA are usually found in many busy urban areas. Different studies, carried out by many authors over the last few decades, have shown that acoustic pollution clearly affects people's health, producing a large series of adverse effects, including noise-induced hearing loss, a reduced ability to communicate, and effects on sleep, work related performance and social behavior.

Although apparently simple, noise pollution is an extensive and complicated subject, with different scientific, technical, economical, political and social aspects, and important repercussions for society. The control of environmental noise requires the efforts of many professionals in fields as different as physics, engineering, architecture, urbanism, health, education, sociology, law and psychology.

This volume constitutes an up-to-date review of our knowledge of environmental urban noise, whilst including the fundamental aspects required to understand this field. Chapter 1 offers a general introduction to the subject. Chapter 2 provides an overview on the physical properties of sound, ratings and descriptors for assessment of environmental noise, and the methods for measuring urban noise levels. Chapter 3 reviews the most important effects of noise on human health, such as sleep disturbance, physiological effects and stress. Chapter 4 refers specifically to the community response to environmental noise, stressing the evaluation of attitudes and subjective responses of residents to urban noise. Chapter 5 provides an introduction to transportation noise sources, outdoor sound propagation and the prediction of urban noise levels produced by road traffic. Finally, Chapter 6 presents the most relevant aspects of urban noise control, from a multidisciplinary point of view (technical, administrative and economic).

The book is designed to be an answer to the need for a comprehensive and up-to-date reference book on this most important field of environmental sciences. Contributions on all topics of environmental noise can certainly be found in many journals, periodicals and conference proceedings. However, most of these articles are written only for a selected number of specialists. Considering the continuing expansion of scientific knowledge on environmental noise, it is very difficult for many readers to obtain a general overview of the most important noise pollution topics without considerable effort.

We hope that *Environmental Urban Noise* will help to solve this problem. Accordingly, all authors have presented their work in an easily accessible form for a broad spectra of readers, whether specialists or otherwise (researchers, teachers, acoustical consultants, postgraduate students, etc.).

I have very much enjoyed planning, editing and writing contributions for this book. I am greatly indebted to all the authors for the very fine chapters they have contributed. I am also grateful for the support and assistance received from Mr. Lance Sucharov and all members of the WIT Press Advances in Ecological Sciences Editorial Board.

Amando García
Valencia, 2001.

Chapter 1

Introduction

Amando García
Department of Applied Physics, University of Valencia, Valencia, Spain

1 Noise pollution

Deterioration of the environment is one of the most serious problems that the humanity must confront nowadays. Among many other effects, the uncontrolled development of human activities has resulted in the pollution of air and water, the hot-house effect, the despoliation of extensive forest areas, the desertization of many lands and the accumulation of huge quantities of toxic wastes. Within this context, modern societies are also experiencing a very important increase in the levels of acoustic pollution. Environmental noise has become one of the greatest sources of nuisance in all developed countries [1][2][3][4].

Noise fills everything and affects everybody. People are constantly exposed to varying degrees of high noise levels in their quotidian activities, for instance, when they are working in an industry, shop or office (occupational noise), using any transportation system (private cars, buses, railways or aircraft), resting at home (noise produced by a wide variety of both external and internal sources), or just being entertained in some leisure places (bars, discotheques or sporting stadiums). For any person that lives in an industrialized society, noise pollution is an absolutely familiar element, something that he/she must learn to live with, although it could be occasionally the target of some complaints or adverse comments. At present, most urban residents are convinced that noise is a very difficult environmental factor to control and, in the last instance, an unavoidable sequel of the technological progress of modern societies.

The different studies carried out by many authors during the last decades have shown that acoustical pollution clearly affects people's health, producing a large series of physical and psychological effects, the impact depending largely on the conditions existing in each case [5][6]. From a non physiological perspective, it should be noticed that all of us can be affected by the perception of a given sound and not by another sound of similar characteristics to the first one. In other words, the magnitude of the nuisance that produces a given noise not only depends on the degree to which the considered noise exceeds the background sound level persons are exposed to in a given location and time, but also of the specific activity that they are performing in that precise instant (reading, conversing, watching TV, sleeping, etc.).

These last considerations are related with the definition of noise as an "unwanted sound" by the receptor or as an "unpleasant and annoying auditive sensation". The imprecise character of these definitions springs from the subjectivity with which certain elements of a specific sound are usually judged (high energy levels, significant variations of intensity with time, etc.). In other words, the degree of annoyance produced by a given sound phenomenon is precisely that which qualifies this sound as noise. Although not treated here, it should be mentioned that environmental noise also affects natural wildlife and ecological systems.

A well known report of the Organization for Economic Cooperation and Development indicated that the dramatic growth of different transportation systems (road traffic, railways and aircraft) in the 1960s and 1970s produced a significant increase in the environmental noise levels in all developed countries. In particular, it has been estimated that about 15-20% of the population in developed countries, i.e., over 100 million people, are actually exposed to diurnal equivalent sound levels (Leq) exceeding 65 dBA [3][7]. The Leq descriptor is the level of a steady sound which, over the same interval of time as the fluctuating sound of interest, has the same mean square sound pressure, usually applied as a normalised A-frequency weighting [8][9].

Sound levels found in our everyday lives show a wide range of values. For instance, a sound level of 130 dBA is considered absolutely unbearable, producing an acute pain sensation in ones ears; this value could be produced by the taking-off of a jet aircraft or by some fireworks, with the observer situated at a distance of 10 metres from the noise sources. A continued exposure to sound levels of 100 dBA can represent a serious risk for human health (hearing impairment); these levels are occasionally reached in some industrial areas. Sound levels of 70-80 dBA are typically found in very busy urban conurbations (main roads with very intense traffic). A sound level of about 50 dBA could correspond to the diurnal background noise level in a quiet urban area. A sound level as low as 20 dBA, actually exceptional, could be found only in some natural open spaces (such as deserts or snow lands quite distant from any activity) or in very special locations (such as an audio recording studio). The threshold of hearing corresponds to a sound pressure level of 20 µPa [8][9].

2 Urban noise sources

Environmental noise has always existed in the world. Mankind has been ever continually exposed to a large range of audible sounds extremely varied both in their origins and physical characteristics. Nature contains an extensive series of noise sources. Some of these natural sources (volcanic eruptions, hurricanes, storms, avalanches, etc.) produce high intensity sounds. Nevertheless, the most aggressive environments are directly related to human activity and appear generally where this activity is more intense and concentrated.

The presence of many noise sources in the cities has been mentioned by some classical writers. In this sense, the references of the Spanish/Latin poet Marcial are quite significant. In one of his *Epigrams*, Marcial referred to the especially noisy character of ancient Rome, describing how the noise produced by the teacher and students of a nearby school prevented his sleeping during the day. When this noise ended, the work of the bakers was responsible for keeping him awake. The tinkers produced a hellish noise as well. This poet alluded also to the screams of beggars and to the strokes of bankers counting their coins with the aim of attracting possible clients. Pliny the Elder, another Roman poet, had a bedroom built with double walls to protect his sleep from the noise generated by his slaves, or coming from the outside of his dwelling. In what is considered as a very early precedent of the modern noise control regulations, it is known that in ancient Rome the traffic of carriages was forbidden during night hours in order to protect the sleep of the residents.

In his *Divine Comedy*, Dante considered that noise was a devil's invention, imagining that some of the damned in hell would be exposed to an unbearable noise, as an endless punishment for their many sins. Making use of a much more realistic approach, some Mongol emperors of the Liao dynasty tortured war prisoners by placing them under huge bronze bells, whose continued percussion finally caused their death. From another perspective, the *Oxford English Dictionary* includes many references on noise as an unwanted sound dating back to the Middle Ages.

However, the most important deterioration of environmental acoustical conditions started with the first Industrial Revolution. Since then, the situation has been made substantially worse, as a consequence of industrial development, the massive growth of modern transportation systems and the emergence of large urban agglomerations. At present, the main sound objects conforming the acoustic ambience in urban zones are related to the transportation systems of people and goods [10]. This category of sound sources is basically formed by three types of vehicles: motor vehicles, aircraft and railways.

The road traffic (cars, buses, lorries and motorcycles) is, by far, the most important and generalized noise source in all urban areas of industrialized countries. This assertion is based both on the results of the corresponding noise level measurements and on the intensity of disturbance that this source produces on urban residents [11]. The noise produced by road vehicles is mainly generated from the engine and from contact between the vehicle and the ground. The study carried out in the United Kingdom in the 1960s by the Wilson Committee, considered as the first serious attempt on a global basis to tackling the problem of environmental noise, arrived at the conclusion that traffic noise was the predominant source of annoyance in urban zones [1]. In the decades since then, the importance of road traffic as the main noise source in all urban areas has increased considerably. The noise pollution levels produced by road traffic usually reach very high values in the main urban or interurban roads, which, in many cases, support very high traffic densities both day and night. In these cases, the noise impact on people living close to these roads can be so intolerable that it makes absolutely necessary the adoption of special protection measures.

In order of importance, the second place of the noise sources related to transportation systems corresponds to aircraft [12]. Aircraft operations have caused severe noise problems since the late 1950s, when turbojet aircraft entered into service. The increase of the number of flights (both civil and military) and the generalized use of this type of transport for freight movement during the last decades has produced a marked increase in air traffic. At present, millions of people around the world are exposed to some noise from aircraft. Obviously, the nuisance from this noise source is most intense in the vicinity of airports, where many aircraft converge flying at relatively low altitudes during their arrival or departure operations.

The third of the noise sources related to transportation are the railways, both above and below ground [11]. In this case, the most important sound emission is produced in the moving units and in the wheel-rail interaction. The levels of sound emission depend basically on the velocity of the trains, but variations are present depending upon the type of engine, wagons and rails. In particular, the introduction of high-speed trains has created very serious noise problems. Nevertheless, the number of studies carried out to evaluate the noise impact produced by the railways in urban zones is comparatively much lower than that concerning road vehicles and aircraft. The results found out in these works have shown that the response of people affected by the noise produced by railways is not so critical as that produced by the other transportation systems.

A quite important noise source in many urban areas are the industries [13]. Of course, all the big factories (metallurgy, cement, ship building, oil refineries, etc.) are usually located fairly distant from residential areas. However, it frequently happens that, due to the spectacular and disordered growth of many cities, many old industrial zones have been fully absorbed by them, making it difficult, in practice, to distinguish between residential and industrial zones. On the other hand, a large number of small industries and workshops, perfectly integrated into the urban fabric, exist in many of our cities, with close connections to the rest of the community (movements of employees, suppliers and clients, arrival of raw materials and departure of manufactured products, loading and unloading operations, etc.), which can produce a very important direct and indirect sound impact on the neighboring zones.

Although more irregularly distributed and less permanent than other noise sources, public works commissions and building construction are activities that can cause considerable noise emissions [14]. The use of equipment such as compressors, cranes, hammers, welders, loaders, excavators, concrete mixers, etc., produces excessive noise. Consequently, these activities are the object of many complaints from affected residents (it should be noted that construction equipment is often poorly silenced and maintained). In the last few years, the construction equipment industry has made a big effort to reduce the noise emission levels of different machines. The need to reduce noise has been triggered by legislative action as well as customer requests for quieter machinery.

Finally, in all urban areas there also exist a very wide variety of other sound sources, which are generally characterized by their singular and sporadic nature, although, unfortunately, their negative impact is noticed with excessive frequency. This is the case, for instance, with sirens of ambulances, police and fire-tender vehicles, or the acoustic signals produced by many security systems (buildings and vehicle alarms). The presence of these communication sound objects is an important nuisance factor in many cities. Although their objective necessity can be widely recognized, many people consider that, in some cases, the use of these acoustic systems is abusive and superfluous.

In a different sense, some building equipment such as ventilation and air conditioning plants, heat pumps, plumbing systems and lifts, can also contribute to the internal acoustic environment of all our buildings, adding up to the immission noise produced outside. A very special mention must be made of noise from neighbors, which is often one of the main causes of noise complaints, largely related to the inconsiderate use of powered home appliances (such as vacuum cleaners, washing machines, chain saws or lawn mowers), systems for music reproduction, radio and television sets, hobby equipment, domestic animals or excessively noisy social celebrations.

The sound sources related to many leisure activities (which intensities, spectra and space localization can be very different) have a very acute social importance. In this group could be included sound sources as different as the voices of children playing in a park or in a school courtyard, the shouts of thousands of people attending a sporting competition, the music played in an open air pop concert, or the sound originating in activities such as fun fairs, street celebrations, pyrotechnic fires, public demonstrations, etc. [15]. The use of certain powered machines, such as off-road vehicles or motorboats, can also contribute to deteriorating the sound ambience in some previously quiet residential areas. Within this group of noise sources should be specially mentioned the bars, pubs, discotheques and similar leisure sites, that can produce a quite important direct and indirect sound impact on nearby residents. This is a very frequent problem in Spain, where most of these activities are the target of many complaints, especially during weekends, when these sites stay open up to early in the morning and many people remain in the streets, chatting and drinking.

All the above mentioned sound sources, and many more, contribute in some amount to what some authors have called the sound landscape of all our cities, a concept which defines the acoustic characteristics of these spaces in a similar way to that in which form and colors are connected to the visual landscape of the same spaces, and, as happens with this case, imprint them with a specific quality that produce a variety of different and complex sensations and emotions [16]. In other words, the acoustic environment of urban areas does not refer only to noise (unwanted sound), but it could have many other connotations besides the nuisance, related in one or another sense with different perceptive parameters. From this point of view, the sound (as light) should be considered as a most important factor for information and communication of people with the environment in which they live. Different experiments combining linguistic and psychological analyses have shown that urban soundscapes include a complex combination of unpleasant and pleasant sounds. The subjective evaluation of a given sound landscape not only depends on the corresponding sound intensity but also, and perhaps even in greater proportion, on the contained information, on the context in which it is perceived and on the social and cultural meanings attributed to it by different individuals [17].

3 Effects of noise on people

The possibility that environmental noise has a detrimental effect on human health has provided a strong incentive for research in this field of work [5][6]. The pernicious effects of the high levels of noise existing in industrial settings on the auditory system of the workers has been a very well established fact for a long time: the workers in certain professions such as boiler-maker or mining have been known to become deaf after years of work. With industrial development, however, the number of workers exposed to excessive noise levels (for instance, in industries such as textile or metal) has increased significantly. On the other hand, in most developed countries the hearing impairment due to community noise exposure (sociacusis) has also become a quite serious problem. Hearing disability may be assessed in terms of difficulty in understanding acoustic signals and speech. Available data show that there is considerable variation in human sensitivity with respect to hearing impairment. Consequently, the hazardous nature of a noisy environment is usually described in terms of damage risk. Referring to occupational exposure conditions, it is generally believed that the hearing loss risk is negligible at noise exposure levels (Leq) lower than 75 dBA for an 8 hour daily working period [6].

Many studies have demonstrated that urban noise exposure may produce a number of direct adverse effects other than hearing damage [18][19]. These include interference on communication, sleep disturbance effects, psychophysiological effects, effects on performance and general annoyance. Although it is quite clear that continued and high level noise exposure produces hearing loss, the relationship between noise and other health effects is not so evident.

The primary method of communication between humans is speech. The interference of noise with speech communication is a masking process in which environmental noise curtails or prevents the speech perception [5]. The ratio of a given signal level (speech or music) to that of interfering noise determines to what extent the signal can be perceived by a listener. In occupational environments, this interference can produce the failure of workers to hear warning signals, a situation that can even lead to serious accidents. Many measurements indicate that, during relaxed conversations at home, the speech level is about 55 dBA. When the noise levels increase people tend to raise their voices to overcome the masking effects. Consequently, the oral communication between two persons who meet in a noisy street can be quite difficult.

Many social survey data have indicated that sleep disturbance is considered to be a most important environmental noise effect [20]. Exposure to noise can produce disturbances of sleep in terms of difficulty to fall asleep, alterations of sleep pattern and depth, and awakening (primary disturbance effects). Exposure to night-time noise can also induce a number of secondary effects, that is, effects that can be measured in the day after the noise exposure, such as increased fatigue, decreased comfort and reduced performance. Some long term effects on psychosocial wellbeing have been also related to night noise exposure. In order to avoid these negative effects, the equivalent continuous sound pressure level Leq during the sleeping period should not exceed 30-35 dBA.

Many authors have postulated that noise acts as a general stress and as such may activate several physiological systems leading to changes such as increases in blood pressure and heart rate and vasoconstriction [6]. The magnitude and duration of these effects are determined in part by individual susceptibility towards noise, lifestyle customs and general environmental conditions. In that sense, some studies have pointed to the existence of a tendency for blood pressure to be higher among people living in proximity to large airports and on urban locations exposed to high levels of road traffic noise than among non-exposed subjects. However, the results obtained in these scenarios are somewhat contradictory. Further prospective studies with a better control of all significant variables will be necessary in order to determine the relationship between noise exposure and cardiovascular health and to identify the groups of risk, if any, to these effects.

Environmental noise can interfere with complex task performance. It is natural to expect that mental tasks and any work that demands a degree of attention and concentration are susceptible to the adverse effects of noise [5]. Some studies have shown the existence of negative associations between chronic exposure to high noise levels (produced principally by aircraft or road traffic) and shortcomings in reading ability among children. Children with pre-existing speech or language difficulties are especially vulnerable to these harmful effects [6]. It also appears that steady noise has little, if any, effect upon many tasks once it has become familiar; many mechanical or repetitive tasks generally found in factory work would fall into this category. Actually, it has been pointed out that the reception of certain acoustic signals (background music) may expedite the fulfillment of some tasks.

Noise annoyance can be defined as a "feeling of displeasure or a negative attitude associated with the exposure to an unwanted sound" [21][22]. In large cities, annoyance or disturbance from noise exposure may be present in a huge majority of the urban residents. Field studies are the main source of information about the effects of noise on people living in a given community. These studies consist usually in a social survey in which residents answer a number of carefully chosen questions about their reactions to environmental noise (in general, or related to a given noise source), and a physical noise survey in which the corresponding noise levels are measured [23].

Urban noise annoyance produces a number of effects on affected residents. People may feel a variety of negative emotions when exposed to noise, and may report anger, disappointment, unhappiness, anxiety or depression. Some of the noise effects are related to everyday behavior patterns, such as closing of windows to prevent noise immission (even in hot weather), changing the distribution of space in dwelling (for instance, locating the bedrooms farther away from the most exposed façades) or blocking the normal use of open spaces or balconies. Environmental urban noise may also affect the social behavior of exposed people (aggressiveness, unfriendliness, etc.) or change certain social indicators (residential mobility, drug consumption, etc.).

In urban societies, annoyance from noise exposure may be present in a majority of the inhabitants. In terms of the numbers affected, annoyance is much more widespread than other effects caused by a noisy environment. In general, annoyance is affected by the mean sound level through a given period, the highest sound level of a noise event, the number of such events, and the time of the day. The available data on the noise-annoyance relationship, obtained through many field surveys carried out worldwide in the past few years, are not inconsistent with the simple, physically based equivalent energy theory, which is represented by the Leq descriptor, and which in many cases is a fairly acceptable approximation to an extremely complicated situation.

4 Physical assessment

The characterization of the different sources of environmental urban noise and the precise assessment of their impact on the quality of life of exposed people are central elements for the formulation of any noise abatement program. Clearly, such a program must be predicated on a quantitative understanding of the contribution of each of the broad array of noise-producing devices. Most people are aware, at least qualitatively, of the impact of aircraft noise on airport communities, and are also aware of the noise produced by the numerous heavy vehicles present on our roads. But noise from other types of vehicles, construction, industrial operations and home appliances, among many other sources, are also recognized as a problem to various segments of society.

Noise sources may be characterized individually and in the aggregate. In order to assess its relative importance and as a basis for its impact evaluation, it is generally enough to carry out a simple measure of the sound level produced by a given source at a particular distance. For example, by comparing the sound level produced by some home appliances at 1 meter measuring distance, one can tentatively conclude that refrigerators typically generating a noise level of about 40 dBA are likely to be a far less serious problem than vacuum cleaners generating about 70 dBA. Further, noise levels at other distances and in other situations characteristic of personal exposure may be estimated by accounting for changes in level as sound propagates through the air and solid structures.

Noise levels, scales and ratings appearing in the literature are related in the sense that all are concerned with the human assessment of noise. The first step to assess noise in terms of probable subjective reactions is to isolate the noise from other influencing factors and devise a measurement of noise level which correlates adequately with noisiness. This measurement should emulate the ear's variation of sensitivity with frequency (A, B, C or D weighted levels). The second step is to combine sound level with time in some way to give a scale representing the sound level exceeded for a given proportion of time (statistical descriptors) or the result of an integration of level with respect to duration (equivalent sound level Leq). A scale may become a rating or index simply by defining a given time period (for instance, daytime, night-time or 24 hour period), in order to produce a noise evaluation for a particular type of noise in particular circumstances (day/night equivalent sound level Ldn) [9][24].

Characterizing urban noise levels in a more general sense is also of use in assessing noise impact. It should be noted that people tend to respond differently to the noise coming from a distant motorway or a construction site than to a readily identifiable single event such as a passing motorcycle or the operation of a lawn mower by a neighbor. In principle, a complete physical description of the outdoor noise environment at a specific urban location should include its noise level, frequency spectrum, and temporal pattern. When used with the A-weighted filter characteristics, a sound level meter accounts for the frequency and magnitude of this sound by weighting the amplitudes of the various frequencies approximately according to human hearing sensitivity. This information must be completed with the temporal pattern of that A-weighted noise level (dBA), which is most easily observed on a continuous graphic-level recording. In general, these records show a fairly steady noise (background level), upon which is superimposed a number of more or less pronounced peaks associated with the occurrence of discrete single events. In that sense, distinct sounds such as passing cars, aircraft overflights, human voices or barking dogs, among many others, could be considered as intrusive noises.

The effective control of noise pollution in extensive urban areas is generally based on the data provided by the corresponding noise map [25]. Among other significant data (main noise sources, urbanistic configurations, road traffic data, etc.), the noise map of a town should include the results of a wide range of sound level measurements (considering short or long time equivalent sound levels Leq, the statistical distributions of the instantaneous sound levels Li or the corresponding percentile levels Lx), adequately distributed over space and time, in such a way that offer a most complete description of the noise pollution situation of the different urban areas, including both the noisy and the quiet locations. In general, the corresponding measurements cover the entire town (with points distributed over a regular square grid) and are carried out in the daytime. In some instances, this basic information is completed by a study of the time variation of hourly noise levels during the 24 hours of the day (differences between day and night), the seven days of the week (differences between working days and weekends) and even the different seasons of the year (differences between summer and winter) [26]. In effect, the works of urban noise cartography are a most valuable instrument for urban planning in subjects such as road traffic regulation or land use management.

The acquisition of a complete picture of the noise environment existing in a given urban area does not necessarily require the realization of an extensive program of sound level measurements, a procedure which is always onerous and expensive. It should be remembered that the environmental sound levels measured in urban areas depend on the noise source characteristics (power, frequency and directivity) and the sound propagation conditions (distance source-receiver, land topography, screening by obstacles, reflection effects from buildings, etc.). Based on all this information, a wide range of prediction methods have been developed in many countries. At present, these methods are widely used to assess the degree of exposure of residents to different urban noise sources (road traffic, aircraft, railways, industries, etc.), for an existing or projected situation (virtual noise map).

In general, the prediction methods can be categorized in three different groups. First, they take the form of a manual method which permits the calculation of the corresponding noise levels using quite simple formulae or tables (without the need for a computer); obviously, this method provides valid results only when applied to a limited number of simple situations. The second method is entirely computer based and it enables the user to carry out a great deal of detailed calculations; this method also enables one to obtain very detailed noise maps of the considered zones (including pictorial representations of the respective noise levels). The third possibility refers to scale-model techniques; this method is suitable in especially complex situations, where the theoretical models are unable to adequately account for the complicated relations among many noise sources and a given urban area [27].

5 Urban noise control

Around the turn of the century, the famous German physician Robert Koch stated that "the day will come when mankind will have to fight noise just as vehemently as cholera and pestilency". Unfortunately, this phrase has not lost its relevance one hundred years later [28]. For instance, despite the considerable efforts devoted to controlling environmental noise, about 25% of the European population is exposed, in one way or another, to daytime transportation noise levels exceeding 65 dBA [7]. The OECD has affirmed that the importance of this problem may probably increase in the future due mainly to an ever- expanding use of numerous and powerful sources of noise (road vehicles, aircraft, construction machinery, etc.), enlarging geographical dispersion of noise sources (together with a greater individual mobility) and increasing spread of noise over time (particularly in the early morning, evenings and weekends) [3]. Many millions of residents in urban areas throughout the world consider now that environmental noise abatement should be treated as a priority subject by responsible administrations.

The control of environmental noise is based on a wide variety of strategies. The most efficient action against excessive noise is always the reduction of noise at source. In that sense, in many developed countries there are regulations on permissible noise emissions from noise sources such as motor vehicles, aircraft, construction vehicles or household and garden equipment. Usually the most effective approach is to redesign or replace noisy vehicles or equipment. It should be remembered that the mechanics of sound wave generation in noise sources may differ in two main categories. One is surface motion of a vibrating solid and the other is turbulence in a fluid medium. Consequently, one of the first steps for noise control should be the reduction of forces and/or flow velocity that create the noise generating vibrations [4][29].

A further reduction in noise levels can be obtained by an action as simple as increasing the distance between the receiver (people) and the noise source. For example, the noise impact produced by a projected airport on residents living in an urban area may be reduced if such an airport is located far away from this area. The noise levels produced by a busy motorway on a residential zone may be substantially reduced by adequate planning and a careful urbanistic design of the corresponding area. Sound transmission can also be controlled by the use of partitions or a wide variety of acoustical barriers. An alternative possibility to reducing the immission of external noise inside dwellings is to provide them with high insulation façades [30].

The administrative measures to reduce noise involve five different scenarios: planning, regulating, enforcement, incentives and investment. The planning involves decisions on the use of resources, guidance and co-ordination oriented to noise control. Regulating defines the rules governing the game, established through the relevant legislation or normatives. Enforcement of existing noise regulations must be made by adequate supervision actions, intended to ensure the compliance of these regulations. Incentives usually include a wide range of both economic and non-economic measures, established to persuade public or private groups concerned. Investment refers to allocation of public funds destined to improve the infrastructure, equipment or research directed to controlling environmental noise [30].

The guidelines applied to control the noise produced by a specific noise source depend on the characteristics of such source. For instance, the reduction of road traffic can be achieved both through reduction of noise emission at the source (development of quieter vehicles and adequate road traffic management) and by control of sound transmission and reception (with measures such as adequate design of the roads, land-use planning, construction of roadside noise barriers or insulation of buildings in residential areas). Of course, the policies applied in this sense vary considerably from one country to another [31].

6 Outline of the book

This book constitutes an up-to-date review of our knowledge on the subject of environmental urban noise. This is a very extensive and complicated matter, including many topics and a broad school of science. Consequently, the book includes only the most important aspects of this field of work. Chapter 2 provides a general overview on the ratings and descriptors for assessment of environmental noise and the methods for measuring urban noise levels (instrumentation, general procedures, spatial and temporal sampling, etc.). Chapter 3 reviews the main effects of noise on health (sleep disturbance, physiological effects, stress, etc.). Chapter 4 refers specifically to community effects of noise (evaluation of attitude and subjective response of residents to the urban noise). Chapter 5 provides an introduction to transportation noise and to the prediction of urban noise (a subject that has rapidly developed these last years). Finally, Chapter 6 presents the main aspects of urban noise control, from a multidisciplinary point of view (technical, administrative and economic). All contributors have presented their material in a clear form, making it easily accessible to a broad spectrum of readers (researchers, teachers, acoustic consultants, administration managers, postgraduate students, etc.), specialists or non-specialists in this field of work.

References

[1]   Wilson Committee, *Noise. Final Report*, Her Majesty's Stationery Office, London, 1976.

[2]   Lara, A. & Stephens, R.W.B. (eds.), *Noise Pollution*, John Wiley and Sons, New York, 1986.

[3]   O.E.C.D., *Fighting Noise in the 1990's*, Report of Organization for Economic Cooperation and Development, Paris, 1991.

[4]   Harris, C.M. (ed.), *Handbook of Acoustical Measurements and Noise Control*, McGraw-Hill, New York, 1991.

[5]   Kryter, K.D., *The Effects of Noise on Man*, Academic Press, Orlando, 1985.

[6]   Berglund B. & Lindvall, T., *Community Noise*, Document prepared for the World Health Organization, Stockholm University, Stockholm, 1995.

[7]   Lambert, J. & Vallet, M., *Study Related to the Preparation of a Communication on a Future EC Noise Policy*, INRETS report no. 9420, Bron Cedex, France, 1994.

[8]   Cuniff, P.F., *Environmental Noise Pollution*, John Wiley and Sons, New York, 1977.

[9]   Crocker, M.J., Rating measures, descriptors, criteria, and procedures for determining human response to noise, *Handbook of Acoustics*, John Wiley & Sons, New York, pp. 775-797, 1998.

[10] Nelson, P.M., Introduction to Transport Noise, *Transportation Noise. Reference Book*, Butterworths, London, pp. 1/1-1/14, 1987.

[11] Hickling, R., Surface Transportation Noise, *Handbook of Acoustics*, John Wiley and Sons, New York, pp. 897-905, 1998.

[12] Eldred, K.M., Airport Noise, *Handbook of Acoustics*, John Wiley and Sons, New York, pp. 883-896, 1998.

[13] Bell, L.H. & Bell, D.H., *Industrial Noise Control*, Marcel Dekker, Inc., New York, 1994.

[14] Braunschweig, G.V., Noise of public works, *Proc. Conference on Noise in Metropolitan Cities,* Spanish Acoustical Society, Madrid, pp. 171-178, 1991.

[15] Axelson, A., Recreational exposure to noise and its effects, *Noise Control Eng. Journal*, vol. 44, pp. 127-134, 1996.

[16] Schaffer, R.M., *The Tuning of the World*, The Canadian Publishers, Toronto, 1977.

[17] Dubois, D. & David, S., A cognitive approach of urban landscapes, Proc. Forum Acusticum 99, Berlin, *Acustica / Acta Acustica*, vol. 85, Suppl. 1, S355, 1999.

[18] Langdon, F.J., Noise and man, *Road Traffic Noise*, Applied Science Publishers, London, pp. 1-26, 1975.

[19] Taylor, S.M. & Wilkins, P.A., Health Effects, *Transportation Noise. Reference Book*, Butterworths, London, pp. 4/1-4/12, 1987.

[20] Vallet, M., Sleep Disturbance, *Transportation Noise. Reference Book*, Butterworths, London, pp. 5/1-5/18, 1987.

[21] Fields, J.M. & Hall, F.L., Community Effects of Noise, *Transportation Noise. Reference Book*, Butterworths, London, pp. 3/1-3/27, 1987.

[22] Fidell, S. & Pearsons, K.S., Community response to environmental noise, *Handbook of Acoustics*, Wiley, New York, pp. 907-915, 1998.

[23] Fields, J.M., *An Updated Catalogue of 318 Social Surveys of Residents' Reactions to Environmental Noise (1943-1989)*, NASA Contractor Report 187553, 1991.

[24] Ford, R.D., Physical Assessment of Transportation Noise, *Transportation Noise. Reference Book*, Butterworths, London, pp. 2/1-2/25, 1987.

[25] Brown, A.L. & Lam, K.C., Urban noise levels, *Applied Acoustics*, vol. 20, pp. 23-35, 1987.

[26] García, A. & Garrigues, J.V., 24-hour continuous sound level measurements conducted in Spanish urban areas, *Noise Control Eng. Journal*, vol. 46(4), pp. 159-166, 1998.

[27] Favre, B.M., Factors Affecting Traffic Noise, and Methods of Prediction, *Transportation Noise. Reference Book*, Butterworths, London, pp. 10/1-10/24, 1987.

[28] Verdan, G., Public aspects of noise control, *Proc. 6th International FASE Congress,* Swiss Acoustical Society, Zurich, pp. 249-258, 1992.

[29] Crocker, M.J., Generation of noise in machinery, its control, and the identification of noise sources, *Handbook of Acoustics*, John Wiley and Sons, New York, pp. 815-848, 1998.

[30] Buna, B. & Burguess, M., Methods of Controlling Traffic Noise Impact, *Transportation Noise. Reference Book*, Butterworths, London, pp. 11/1-11/26, 1987.

[31] European Commission Green Paper, Future Noise Policy, *Noise/News International*, vol. 5, pp. 77-98, 1997.

Chapter 2

Physical assessment and rating of urban noise

Giovanni Brambilla
CNR-Istituto di Acustica "O.M. Corbino", Roma, Italy

1 Physical properties of sound[1]-[3]

Sound is the mechanical vibration in a medium (a gas, liquid, or solid) which is detectable by the human ear. This definition points up the twofold characteristic of sound: its physical nature and the psycho-physiological aspect of the hearing sensation.

Sound waves in air are caused by pressure fluctuations above and below the static value of atmospheric pressure (standard value 101,325 Pa), often following a repetitive pattern. The magnitude of these fluctuations is generally in the range from $2 \cdot 10^{-5}$ Pa (threshold of hearing) to 200 Pa (threshold of pain). Sound waves travel through the medium at the local value of sound speed, c, which for the air at 20°C and standard atmospheric pressure is equal to 344 m/s. The speed of sound is independent of pressure fluctuations (at least below 20 Pa) and depends only on the properties of the medium, increasing with its temperature. The more dense and less compressible the medium, the higher the speed of sound in the medium will be.

The number of times per second at which pressure fluctuations oscillate between positive and negative values is denoted as frequency, f, and its unit is called Hertz (Hz). The human hearing is able to detect only the pressure fluctuations within the range of the audible frequencies, that is approximately between 20 Hz and 20,000 Hz. The upper limit generally decreases with increasing age (presbyacusis) and the lower limit is also rather uncertain. Hearing sensitivity is highest around 3000 Hz and decreases both at higher frequencies and quite markedly below about 250 Hz.

The wavelength of sound, λ, is related to the speed of sound, c, and the frequency, f, through the expression:

$$\lambda = c / f \qquad\qquad \text{m} \qquad (1)$$

This parameter is important in diffraction of sound waves occurring when their wavelength is comparable with the dimensions of obstacles along the sound propagation path (i.e. noise barriers).

Except at very low frequencies, the human brain does not respond to the instantaneous pressure fluctuations, but rather to the corresponding root mean square value (rms) defined as follows:

$$\bar{p} = \left[\frac{1}{T} \int_0^T p^2(\tau)d\tau \right]^{1/2} \qquad \text{Pa} \qquad (2)$$

where T is the time period during which the linear averaging of the instantaneous sound pressure $p(\tau)$ is computed. Exponential averaging, however, is more suitable to represent the human hearing which calculates a running average depending more upon the present and less upon the past pressure fluctuations.

The rms value of sound pressure due to such averaging is given by:

$$\bar{p} = \left[\frac{1}{RC} \int_{-\infty}^t p^2(\tau)e^{(\tau-t)/RC}d\tau \right]^{1/2} \qquad \text{Pa} \qquad (3)$$

where RC is the time constant of the averaging circuit. Sound level meters usually have "Fast (F)" and "Slow (S)" detector circuits with RC equal to 125 ms and 1000 ms respectively[4]. It has been shown[5] that approximate equivalence between linear and exponential averaging is obtained by taking $T \cong 2RC$.

The sound intensity, I, is defined as the acoustic energy flowing through unit area (perpendicular to the direction of propagation of sound wave) in unit time. When a sound wave propagates away from the source in a free space, the sound intensity I in the direction of the wave is proportional to sound pressure according to the following relationship:

$$I = \bar{p}^2 / \rho \cdot c \qquad \text{watt/m}^2 \qquad (4)$$

where ρ is the density of air (1.2 kg/m^3 at standard atmospheric pressure) and c is the speed of sound.

Any sound source has a characteristic sound power, W, measured in watts, which is a basic measure of its acoustic output, depending on the physical property of the source alone. Therefore sound power W is an important absolute parameter widely used for rating and comparing sound sources.

1.1   Levels of acoustic parameters

As the human hearing detects a wide range of pressure fluctuations, around 10^7 in magnitude, the application of linear scales to the measurement of sound pressure would lead to enormous and unwieldy numbers. In addition, the ear responds not linearly but logarithmically to the sound stimulus. For these reasons it has been found more practical to express acoustic parameters as the logarithm to the base ten of the ratio of the measured value to a reference quantity. The resulting unit, called Bel, was found in practice to be rather too large, and the decibel, that is one-tenth of a Bel, is now in general use. The decibel scale gives the level of the acoustic parameter in decibels (dB) above (or below) the reference level that is determined by the reference quantity.

The sound pressure level L_p, usually reported as sound level, is defined as follows:

$$L_p = 10 \cdot \log_{10}\left(\frac{\bar{p}^2}{p_{ref}^2}\right) = 20 \cdot \log_{10}\left(\frac{\bar{p}}{p_{ref}}\right) \qquad \text{dB} \qquad (5)$$

where $p_{ref} = 20$ µPa for airborne sound. As this value is approximately equal to the threshold of hearing at 1000 Hz, the sound pressure level in dB is positive and varies typically between 0 dB and 120 dB, before the sound starts to hurt the ear. According to eqn (5), doubling of any value of sound pressure corresponds to an increase in sound pressure level of 6 dB; a multiplication of sound pressure by a factor of 10 corresponds to an increase in sound pressure level of 20 dB.

The sound power level L_W is defined as:

$$L_W = 10 \cdot \log_{10}\left(\frac{W}{W_{ref}}\right) \qquad \text{dB} \qquad (6)$$

where $W_{ref} = 10^{-12}$ watt, that is 1 pW. Substituting this value in eqn (6) gives:

$$L_W = 10 \cdot \log_{10} W + 120 \qquad \text{dB} \qquad (7)$$

The sound intensity level L_I is defined by:

$$L_I = 10 \cdot \log_{10}\left(\frac{I}{I_{ref}}\right) \qquad \text{dB} \qquad (8)$$

where $I_{ref} = 10^{-12}$ watt/m^2. By choosing this value for reference intensity the values of the L_p and L_I for any free progressive sound wave in air differ by a small quantity, about 0.13 dB, and can be neglected for most noise measurements.

1.2 Sound spectra

The simplest sound wave is a continuous pure tone at a fixed single frequency, as shown in Figure 1(a). Unfortunately this sound rarely occurs in practice and more often variations both in amplitude and frequency content with time are observed. Sounds may be a combination of tones whose frequencies are harmonically related, such as those generated by musical instruments. More complicated repetitive waveforms are reported in Figures 1(b) and 1(c), while a signal typical of those encountered in practice is shown in Figure 1(d): it is non-periodic, contains all possible frequencies in a given range and may be viewed as a combination, however complex, of a large number of superimposed sinusoids.

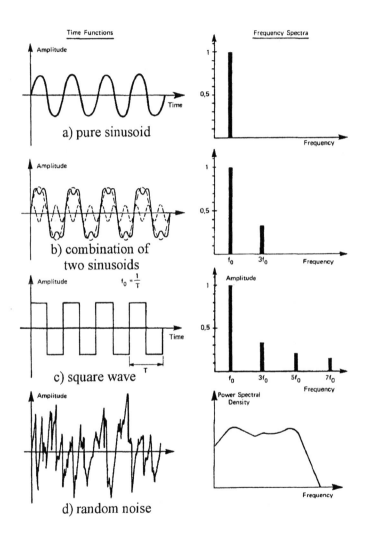

Figure 1: Sound signals and their spectra[3].

To describe the frequency content of such a signal the full frequency range is divided into a series of contiguous bands and the signal level in each band is measured over a sufficiently long time to obtain a meaningful average. The resulting frequency information is known as a spectrum, which can be obtained directly by spectrum analyzers.

Frequencies can be grouped into bands which may be equal in width, as in the Fast Fourier Transform analyzers, or they may have widths proportional to the center frequencies of their respective bands (constant percentage bandwidth filters). The bandwidth is usually chosen to be only as narrow as is necessary to give a spectrum appropriate for the analysis purpose.

Among the constant percentage bandwidth filters octave and 1/3 octave bands are the most frequently used and they are standardized in the center frequencies, shown in Table 1, and bandwidths[6].

Table 1. Octave and 1/3 octave band center frequencies in Hz[6].

Octave Bands	1/3 Octave Bands	Octave Bands	1/3 Octave Bands
	25		800
31.5	31.5	1000	1000
	40		1250
	50		1600
63	63	2000	2000
	80		2500
	100		3150
125	125	4000	4000
	160		5000
	200		6300
250	250	8000	8000
	315		10,000
	400		12,500
500	500	16,000	16,000
	630		20,000

The audible frequency range contains 10 octave bands, each having a bandwidth such that the upper limiting frequency of the passband f_u is twice the lower limiting frequency f_l, resulting in a bandwidth of 70.7%. The center frequency f_c is defined as the geometric mean of f_u and f_l, that is $f_c = \sqrt{f_u \cdot f_l}$.

As the name implies, 1/3 octave bands are formed by dividing each octave band in three parts. The ratio f_u/f_l is equal to $2^{1/3}$, resulting in a percentage bandwidth of 23.1%.

The overall sound level L_{pT} of a signal is obtained from its spectrum by summing the sound level in each frequency band raised to the power of 10, as follows:

$$L_{pT} = 10 \cdot \log_{10}\left(\sum_{i=1}^{n} 10^{L_{pi}/10}\right) \qquad \text{dB} \qquad (9)$$

where n is the number of the frequency bands and L_{pi} is calculated by eqn (5).

Narrow band analysis may be necessary if a sound contains a discrete frequency. In this case the bandwidth is chosen to suit the purpose of the analysis and the result is presented as the spectral density, i.e. the $\overline{p}^2(f)$ per Hz.

1.3 Combining separate sources

Eqn (9) can also be applied to combine the sound levels L_{pi} due to n different sources to obtain the total sound level L_{pT}. The relationship is valid in most of the cases encountered in practice, except when two or more sources are coherent, i.e. the emitted sound waveforms are identical in shape although they may be displaced in time. Coherent sources are unlikely to be encountered in environmental acoustics as they occur when the sources are correlated in some way. An important application of such sources is the active noise control technique, in which a sound wave is generated to be opposite in phase ($\pm 180°$) with respect to the sound wave to control. The latter is reduced in level due to the destructive interference between the two sound waves.

Because of the logarithmic units, the sum of two levels L_{pi} having the same value, for instance 70 dB, is equal to this value increased by 3 dB, that is 73 dB in the example. When the two levels L_{pi} differ by more than 10 dB, for instance $L_{p1} = 70$ dB and $L_{p2} = 59$ dB, the contribution of the lower level to the total level L_{pT} is small, about 0.3 dB in the example, and can be neglected.

1.4 Propagation of sound in the open air[7]

Sound propagating outdoors through the air generally decreases in level with increasing distance between source and receiver. This attenuation is the result of several mechanisms, principally geometric divergence from the sound source, absorption of acoustic energy by the air, the effect of propagation close to ground surface, reflections from buildings, and diffraction around obstacles along the propagation path. Other effects are due to atmospheric conditions, mainly wind and temperature, especially at distances between source and receiver greater than 100 m.

The total attenuation A_T is given by[8]:

$$A_T = A_d + A_a + A_g + A_r + A_s + A_m \qquad \text{dB} \qquad (10)$$

where positive values represent a decrease in sound level. The terms in eqn (10) give the attenuation from geometric divergence A_d, sound absorption by air A_a, ground effect A_g, reflections from buildings A_r (negative values), obstacles along the sound propagation path A_s (i.e. noise barriers), and attenuation from additional mechanisms A_m. In urban environments the most significant attenuations are A_d, A_g, A_r and A_s. The attenuation A_m occurs only in specific cases and includes the attenuations when sound propagates through dense vegetation, industrial sites and blocks of houses.

The terms in eqn (10) should be evaluated separately, as if the others were absent, and then the results added together to give A_T.

1.4.1 Geometric spreading of sound

Many common sound sources, including industrial plant, aircraft, and individual road vehicles, can normally be treated as point sources, i.e. their dimensions are small in relation to their distance from the receiver. When the wavelength of the emitted sound is very large compared with the dimensions of the source, sound is radiated uniformly in all directions, i.e. the source is nondirectional. For such a source located in a free space, and neglecting the air absorption, the entire sound power output W must pass through any notional surface surrounding the source. The sound intensity I at distance r is obtained from W by:

$$ I = \frac{W}{4\pi r^2} \qquad \text{watt/m}^2 \qquad (11) $$

or, in logarithmic form, for any free progressive sound wave in air:

$$ L_p \cong L_I = L_W - 11 - 20 \cdot \log_{10} r \qquad \text{dB} \qquad (12) $$

The final term in eqn (12) represents the inverse square law; according to this law the sound level L_p decreases by 6 dB per doubling of distance from the source.

If the source is directional its behavior is specified by the directivity factor DF, defined as the ratio of the actual sound intensity in a given direction to the sound intensity of an omni-directional source having the same sound power W. Considering the directivity index DI, the logarithmic form of DF, eqn (12) is modified as follows:

$$ L_p \cong L_I = L_W - 11 + DI - 20 \cdot \log_{10} r \qquad \text{dB} \qquad (13) $$

A nondirectional source on a flat reflective surface radiates equally into a hemisphere rather than a sphere and the DI is equal to 3 dB.

A source of great interest is that formed by n point sources equally spaced at distance b along a line on a hard flat surface, as shown in Figure 2. Assuming that the point sources (such as vehicles or trains) are incoherent and nondirectional and each emits the same sound power level L_W, the sound level at the receiver L_p, neglecting the air and ground absorption, is determined by:

$$ L_p = L_W + 10 \cdot \log_{10}\left(\frac{\alpha_n - \alpha_1}{d \cdot b}\right) + \Delta L - 8 \qquad \text{dB} \qquad (14) $$

where $\alpha_n - \alpha_l$ is the aspect angle of the n sources in radians, d is the perpendicular distance from the line to the receiver and ΔL a correction which is less than 1 dB if both the following conditions are met:

$$n \geq 3 \quad \text{and} \quad \frac{d}{b \cdot \cos \alpha_1} \geq \frac{1}{\pi} \qquad (15)$$

When the second condition is met, the number of point sources on the line is equivalent to a continuous line source. In this case the receiver's distance from the closest point source is longer than about one-third the distance b. If the above condition is not met, the receiver is so close to the nearest point source that the L_p level can be calculated by considering exclusively the radiation from this nearest point source with an error of less than 1 dB.

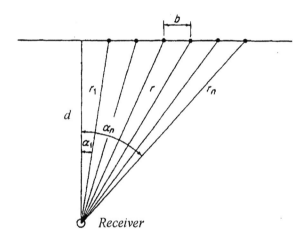

Figure 2: Set of n incoherent point sound sources on a line.

For an infinite number n of point sources $\alpha_n - \alpha_l = \pi$ and from eqn (14) the following relationships are obtained:

$$L_p = L_W - 10 \cdot \log_{10}(d \cdot b) - 3 \quad \text{for } d \geq b/\pi \qquad \text{dB} \qquad (16)$$
$$L_p = L_W - 20 \cdot \log_{10} d - 8 \quad \text{for } d < b/\pi \qquad \text{dB} \qquad (17)$$

Therefore, close to the source ($d < b/\pi$) the sound pressure level decreases by 6 dB per doubling of distance, whilst far from the source ($d > b/\pi$) it decreases at the rate of 3 dB per doubling of distance.

For an infinite number n of point sources within a line of length l, which form an incoherent line source, eqn (14) is generally valid with $\Delta L = 0$ and considering the definitions as in Figure 3:

$$L_p = L_{Wl} + 10 \cdot \log_{10}\left(\frac{\alpha_2 - \alpha_1}{d \cdot l}\right) - 8 \qquad \text{dB} \qquad (18)$$

where L_{Wl} is the sound power level for the entire line source, α_1 and α_2 the angle of the vector to the start and the end of the line source respectively. Close to the line source, where $\alpha_2 - \alpha_1 \to \pi$, eqn (16) can be used instead of eqn (18) with L_W replaced by L_{Wl} and with b replaced by l. Far from the line source the following relationship is obtained:

$$L_p = L_{Wl} - 20 \cdot \log_{10} r - 8 \qquad \text{dB} \qquad (19)$$

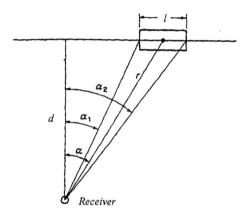

Figure 3: Finite line source, equivalent to a train or a line of vehicles in the case of high traffic flow.

1.4.2 Air absorption

As sound propagates through the atmosphere its energy is gradually converted into heat by molecular processes in the air, resulting in the sound absorption. The attenuation of sound due to air absorption during propagation, A_a, through a distance d (meters) is given by:

$$A_a = \alpha \cdot d / 100 \qquad \text{dB} \qquad (20)$$

where α is the air attenuation coefficient in decibels per kilometer; its value is available in tables and graphs[9]. The sound absorption depends strongly on frequency (increasing with it) and relative humidity (inversely proportional), and less strongly on temperature. It also depends slightly on the ambient pressure. The attenuation A_a can be neglected at short distances from the source (less than a few hundred meters), except for frequencies above 5000 Hz.

1.4.3   Ground attenuation

When both sound source *S* and receiver *R* are close to the ground, as frequently occurs in outdoor sound propagation, the sound reaches the receiver both directly along path r_d and by reflection from the ground along path r_r, as shown in Figure 4. The reflected wave appears to come from the image source *I*, specular of *S* with respect to the ground. The sound attenuation A_g is a result of interference between direct and ground reflected sounds, and strongly depends on the type of ground surface, the grazing angle ψ, the path length difference $r_r - r_d$, and the frequency of the sound.

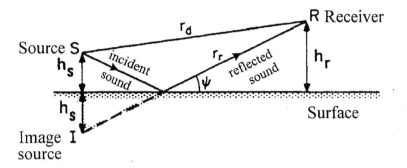

Figure 4: Sound source and receiver in the vicinity of a reflecting ground plane.

For grazing angles less than 20°, covering most of the sound propagation outdoors, the ground surface may broadly be classified in four categories:

- hard ground, including asphalt or concrete pavement, water, and all other ground surfaces having a low porosity;
- soft ground, such as ground covered by grass, trees, or other vegetation, and all other porous grounds suitable for the growth of vegetation;
- very soft ground, such as ground covered with snow, pine needles, or similarly loose material;
- mixed ground, including both hard and soft areas.

At grazing angles greater than about 30°, commonly occurring at short ranges, soft and very soft grounds become good reflectors of sound and, therefore, they should be considered as hard ground.

When the following conditions are met:

- propagation over ground mainly soft;
- sound spectrum broad and smooth without prominent discrete frequency components, as frequently occurs for major noise sources such as industrial plants and road traffic;
- only the A weighted sound pressure level at the receiver is of interest;

then the ground attenuation can be considered independent of the frequency and its calculation may be simplified as follows[8]:

$$A_g = 4.8 - \left(2h_m / r\right) \cdot \left(17 + 300 / r\right) \qquad \text{dB} \qquad (21)$$

where r is the distance between source and receiver and h_m is the mean height of the propagation path above the ground; for flat ground $h_m = (h_S + h_R)/2$. Negative values for A_g obtained from eqn (21) have no significance and should be replaced by zeros.

For conditions different from those described above, alternative calculation procedures for both short and long ranges are reported in the ISO standard 9613 Part 2[8].

1.4.4   Reflections

Sound arriving at the receiver by reflections from a vertical surface, such as a building façade, enhances the sound directly proceeding from source to receiver. For this situation the attenuation due to reflection, A_r, has a negative value and may be calculated in the same way as for A_g, taking into account that normally the surface is acoustically hard.

1.4.5   Barriers

A sound barrier is any solid obstacle which is relatively opaque to sound and blocks the line of sight from sound source to receiver, as shown in Figure 5. Barriers may occur naturally, such as earth berms and buildings, or they may be installed specifically to reduce noise.

The usual measure of the acoustical effectiveness of a barrier is its insertion loss D_{IL}, defined as the difference, for a given point, of the sound levels measured before and after the installation of the barrier. It can be determined for a single frequency, a band of frequencies or all the frequency range. This measure avoids the ambiguity that arises because the barrier, besides introducing attenuation due to diffraction as shown further on, also commonly reduces the attenuation due to the ground (by increasing the height of the propagation path above the ground). The insertion loss depends on several parameters, most notably the frequency of the sound (the higher frequencies are more attenuated). The construction of the barrier must also ensure that the noise is not able to pass through the barrier. For this purpose large cracks or gaps should be avoided in the barrier and this should have a superficial mass of at least 10 kg/m^2.

The insertion loss by an infinitely long thin barrier, attenuating by single diffraction the sound with wavelength λ from a point source, is calculated from the following relationship:

$$D_{IL} = 10 \cdot \log_{10}\left(3 + 10NK\right) - A_g \qquad \text{dB} \qquad (22)$$

where N is the Fresnel number given by:

$$N = \frac{2}{\lambda} \cdot \left(d_1 + d_2 - d\right) \qquad (23)$$

where d_1, d_2 and d are the distances shown in Figure 5. When the tip of the barrier just touches the line of sight between source and receiver the value of N is zero and $D_{IL} \cong 5$ dB. Positive values of N are obtained when the receiver is the shadow zone provided by the barrier, whilst negative values correspond to the receiver not shielded by the barrier (bright zone).

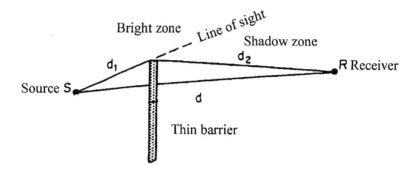

Figure 5: Geometry of a thin barrier.

In eqn (22) the term A_g is the ground attenuation before the barrier is installed, while the first term is the attenuation provided by the barrier plus any attenuation still effective in the propagation path resulting from the ground and atmospheric effects after the installation of the barrier. The correction factor K for atmospheric effects in eqn (22) is set to 1 for distances between the source and receiver less than 100 m. For distances between 100 m and 300 m the reduction of D_{IL} due to the atmospheric effects increases with the distance and the value of K which is given by[8]:

$$K = \exp\left[-0.0005 \cdot \sqrt{(d_1 d_2 d)/(N\lambda)}\right] \tag{24}$$

Negative values of insertion loss from eqn (22) must be set to zero.

A thick barrier, such as a building or an earth berm, attenuates the sound by double diffraction, as shown in Figure 6. If the barrier thickness, t, is greater than 3 m, the barrier is considered as thick for all the frequencies. If t is less than 3 m, the barrier is still regarded as thick for wavelengths less than $t/5$, otherwise the barrier should be considered as thin and its insertion loss calculated by eqn (22).

For a thick barrier the Fresnel number N is given by:

$$N = \frac{2}{\lambda} \cdot (d_1 + t + d_2 - d) \tag{25}$$

where d_1, d_2, d and t are the distances shown in Figure 6. Then the insertion loss D_{IL} of the barrier is calculated by[8]:

$$D_{IL} = 10 \cdot \log_{10}(3 + 30NK) - A_g \qquad \text{dB} \qquad (26)$$

where A_g and K have the same meanings as in eqn (22). In calculating the value of K by eqn (24) the thickness t has to be added to the smaller of the two distances d_1 or d_2. Negative values for D_{IL} from eqn (26) must be replaced by zero.

The thick barrier can also be replaced by a notional thin barrier having the equivalent height h_e shown in Figure 6.

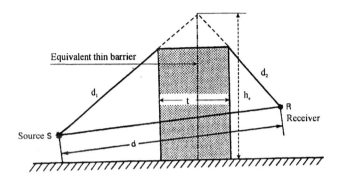

Figure 6: Geometry of a thick barrier.

For a barrier of finite length, three propagation paths need to be considered between source and receiver, as shown in Figure 7: path *a* over the top of the barrier, and paths *b* and *c* around each end. The level at the receiver is obtained as the sum of the levels, according to eqn (9), calculated separately for each path. The distances d_1, d_2, d and t are determined considering the relevant plan view of each propagation path; for paths *b* and *c* set ground attenuation $A_g = 0$ and $K = 1$.

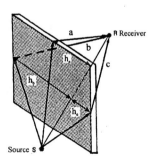

Figure 7: Sound paths around a barrier of finite length.

1.4.6   Meteorological conditions

The sound propagation close to the ground for distances between source and receiver of less than about 100 m is essentially independent of atmospheric conditions; for this case the atmosphere can be regarded as homogeneous and the ray paths approximated by straight lines, as shown in Figure 4. For greater distances, atmospheric conditions usually become a major factor and the main effect is refraction, a change in the direction of the sound waves, produced by vertical gradients of wind and temperature.

The wind speed usually increases with height above the ground. Consequently, in the downwind direction the top side of the wavefront is pushed forward and the sound rays curve down towards the ground, as shown in Figure 8. On the contrary, upwind the rays curve upwards away from the ground creating a shadow zone, as illustrated in Figure 8.

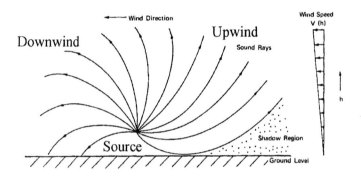

Figure 8: Sound ray curvature due to wind speed gradient.

Temperature gradients produce a similar effect because, to a first approximation, the rate of change of sound speed with height is proportional to the temperature gradient. Consequently, in lapse temperature conditions, which generally occur during the day when the ground is warmest, the air temperature and the sound speed decrease with increasing height and the sound rays curve upwards. This creates a shadow zone all around the source unlike that arising from a wind gradient, as shown in Figure 9(a). In contrast, in inversion conditions, which tend to occur in the late evening or at night when skies are clear and the ground has cooled rapidly, there is a region above the ground, which may extend to 100 m or more, where the air temperature, and therefore the sound speed, increases with height and the rays bend down towards the ground, so enhancing the sound level, as shown in Figure 9(b).

Upward refracting conditions (i.e. upwind and temperature lapse) produce a shadow zone near the ground, resulting in an extra attenuation that typically reaches 20 dB or more.

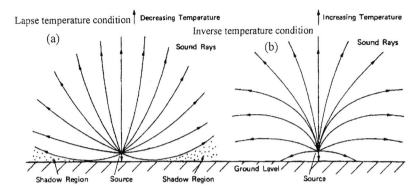

Figure 9: Sound ray curvature due to temperature gradient.

On the contrary, downward refracting conditions (i.e. downwind and temperature inversions) are favorable for sound propagation and produce a minimum attenuation. However, they are stable and, therefore, suitable for reproducible measurements, being the attenuation relatively insensitive to minor changes of atmospheric conditions. For these reasons it has become standard practice to consider only atmospheric conditions favorable for sound propagation in the description of environmental noise. However, sound levels determined over long periods, ranging from a month to a year, are usually recommended. This period may include a great variety of meteorological conditions, some of which are less favorable for sound propagation.

2 Scales and ratings[1], [10]

Noise is defined as any unwanted sound that is undesired to be heard by the subject for several reasons, such as its interference with an activity undertaken by the subject, sleep disturbance, time of occurrence and so on.

Therefore all efforts to reduce noise pollution rest on the means for describing the magnitude of sound as it affects human beings. This aim demands the development of a numerical evaluation of sound, preferably in terms of a single number, bearing a meaningful relationship to the human response to noise. Thus, we need to look for ways to measure certain physical properties of sound that are closely connected with people's subjective judgment.

Assessing noise in terms of probable subjective reactions requires three steps, each of which should be validated by psychoacoustic experiments, usually performed in the laboratory, and social surveys undertaken in the field.

The first step is to isolate the noise from other influencing factors and devise a measurement of level which correlates with noisiness, that is the degree of unwantedness of a sound. Such a measurement should emulate the ear's variation of sensitivity with frequency and, possibly, take masking into account. At this stage time is not included so the level is function of time. The "A", "B", "C" and

"D" weighted levels fall into this category. All these parameters are usually denoted with the term "unit".

The second step is to combine level with time in some way to give a "scale". This may be the level exceeded for a given proportion of the time as in the statistical centile level or it may be an integration of level with respect to duration as in the equivalent continuous level and single event noise exposure.

The "rating" or "index" goes one step further, where the time period is clearly specified and other factors are considered, which may affect people's reactions according to the circumstances or time at which the noise is heard. An "index" may be considered as an adjusted scale to be used as a basis for assessment in planning and regulations. Day/night level, day or night and 24-hour equivalent continuous level are examples of "index".

The "units", "scales" and "indexes" most relevant and frequently used in the environmental urban noise assessment will be described in the following.

2.1 Loudness

Loudness is defined as the subjective evaluation of the intensity of a sound and depends not only on its sound level but also on its frequency, as shown by the equal loudness contours in Figure 10. The contours connect together pure tones which are judged equally loud in free field listening conditions by young adults with normal hearing, facing the source. The dashed contour at 4.2 phons represents the minimum audible field (MAF).

The unit of loudness is the sone. A sone is defined as the loudness of a pure tone at 1000 Hz having a sound level of 40 dB. A sound that is twice as loud has a loudness of 2 sones and so on. For an average listener a 10 dB change in sound level is roughly equivalent to a doubling of loudness.

The loudness level P of a 1000 Hz pure tone of 40 dB is defined as 40 phons and the relationship between loudness S and loudness level P is given by:

$$S = 2^{(P-40)/10} \quad \text{sone} \qquad P = 40 + 10 \cdot \log_2 S \quad \text{phone} \qquad (27)$$

Thus, a doubling of loudness in sones is equivalent to an increase of 10 phons in loudness level.

Equal loudness contours for bands of noise have been determined experimentally by Stevens[12] and Zwicker[13] and can be used to evaluate the loudness of noise sources. The procedures developed by Stevens and Zwicker take into account that the ear is able to mask sounds close to each other in their frequency. Both methods are internationally standardized[14] and can only be applied to steady sounds.

The Stevens Mark VI method assumes that the sound field is diffuse and does not contain any prominent pure tones. For each of the n frequency bands of the noise spectrum (preferably 1/3 octave bands) a loudness index S_i is determined from the set of equal loudness index contours plotted in Figure 11. The total loudness S is then given by:

$$S = S_{max} + F\left(\sum_{i=1}^{n} S_i - S_{max}\right) \qquad \text{sone} \qquad (28)$$

where S_{max} is the maximum loudness index, $F = 0.3$ for octave bands and $F = 0.15$ for 1/3 octave bands.

Stevens has also proposed a Mark VII procedure[15] which appears to be more accurate and extends loudness calculations to both low sound pressure levels and lower frequencies and high sound pressure levels.

The Zwicker method is based on the critical band concept. A critical band is the largest frequency bandwidth of flat (constant spectrum level) random noise having the same loudness as a pure tone of the same sound level at a frequency equal to the geometric mean frequency of the critical band. This method, although more complicated than Stevens' procedure, can be used for sound having prominent tonal components and either in diffuse sound field or frontally incident sound in a free field. The 1/3 octave band spectrum is required, and complete details of the procedure are given in the ISO standard[14].

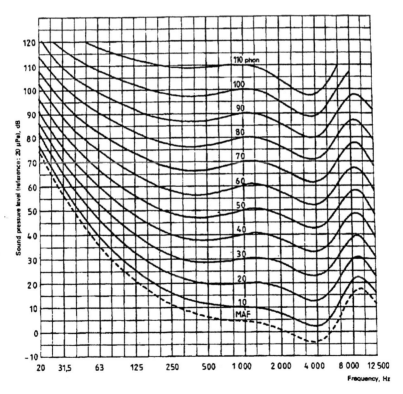

Figure 10: Equal loudness contours for pure tones[11].

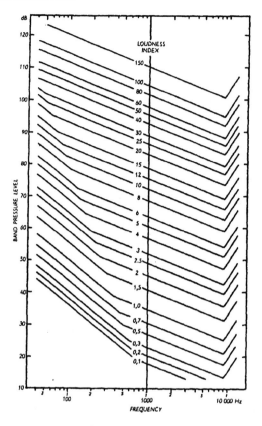

Figure 11: Equal loudness index contours[14].

If the sound has short duration (less than about 200 ms), its loudness is reduced compared with the same sound heard continuously. The shorter the duration becomes, the less loud the sound appears to be.

2.2 Weighted sound levels

One approach to measuring sounds in a way reflecting on their loudness is to alter the measured spectrum of the sound to take account of the fact that the sensitivity of human hearing is dependent on frequency. Thus, the resulting weighted sound pressure level should give some indication of the loudness of the sound. Three such weightings were originally proposed and now are internationally standardized, namely the "A", "B" and "C" weighting curves shown in Figure 12. The "A" weighting filter was designed to approximate the inverse of the equal loudness contour at 40 phons and should be applied for sounds not loud, having loudness levels below 55 phons. The "B" and "C" weighting filters were designed to follow approximately the inverse of the equal loudness contour at 70 and 100 phons respectively and should be applied for sounds moderately loud ("B" curve, loudness levels between 55 and 85 phons) and loud ("C" curve, loudness levels over 85 phons).

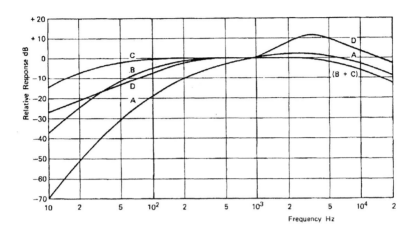

Figure 12: Weighting curves standardized for sound level meters.

To specify the weighting applied the sound pressure level is given in units of dB(A), dB(B) or dB(C). The specification for each of the three weightings is shown in Table 2 for 1/3 octave bands and octave bands (typed in boldface). The table also contains the "D" curve specification; this weighting, plotted in Figure 12, has been introduced for measuring aircraft noise, to account for the increase in annoyance produced by the high frequency whine present in such a noise.

The A-weighted sound level has been shown to correlate reasonably well with the subjective response on loudness of broad-band sounds and even on the acceptability of the noise. This performance, together with its easy implementation on sound level meters, has led to a wide use of the A weighting, which is now extensively applied to all levels of noise and adopted in many national and international legislations and standards. In addition, the A-weighted sound level forms the basis of many other descriptors.

The B weighting is seldom used because it offers no advantages over the A weighting. The C weighting differs little from a flat frequency response and it is generally used both to limit the low- and high-frequency response of the instrument and to measure high-energy impulsive noises, such as a sonic boom.

Table 2. The A, B, C and D weightings

1/3 Octave Band Centre Frequency Hz	Correction dB			
	A	B	C	D
25	-44.7	-20.4	-4.4	-18.7
31.5	**-39.4**	**-17.1**	**-3.0**	**-16.7**
40	-34.6	-14.2	-2.0	-14.7
50	-30.2	-11.6	-1.3	-12.8
63	**-26.2**	**-9.3**	**-0.8**	**-10.9**
80	-22.5	-7.4	-0.5	-9.0
100	-19.1	-5.6	-0.3	-7.2
125	**-16.1**	**-4.2**	**-0.2**	**-5.5**
160	-13.4	-3.0	-0.1	-4.0
200	-10.9	-2.0	0.0	-2.6
250	**-8.6**	**-1.3**	**0.0**	**-1.6**
315	-6.6	-0.8	0.0	-0.8
400	-4.8	-0.5	0.0	-0.4
500	**-3.2**	**-0.3**	**0.0**	**-0.3**
630	-1.9	-0.1	0.0	-0.5
800	-0.8	0.0	0.0	-0.6
1000	**0.0**	**0.0**	**0.0**	**0.0**
1250	0.6	0.0	0.0	2.0
1600	1.0	0.0	-0.1	4.9
2000	**1.2**	**-0.1**	**-0.2**	**7.9**
2500	1.3	-0.2	-0.3	10.4
3150	1.2	-0.4	-0.5	11.6
4000	**1.0**	**-0.7**	**-0.8**	**11.1**
5000	0.5	-1.2	-1.3	9.6
6300	-0.1	-1.9	-2.0	7.6
8000	**-1.1**	**-2.9**	**-3.0**	**5.5**
10,000	-2.5	-4.3	-4.4	3.4
12,500	-4.3	-6.1	-6.2	1.4
16,000	**-6.6**	**-8.4**	**-8.5**	**-0.7**
20,000	-9.3	-11.1	-11.2	-2.7

2.3 Statistical centile sound level

Environmental urban noise typically shows sound levels varying with time, as illustrated in the example given in Figure 13(a). To try to account for fluctuations in noise level and the intermittent character of some noises (such as passing road vehicles or aircraft movements), statistical centile sound levels are frequently used. The statistical centile sound level, L_n, is defined as the level, usually expressed in dB(A) and measured by the "Fast" time constant, exceeded for the n percentage of the measuring time; for example L_{10} represents the sound level exceeded for the 10% of the time. The values of L_n can be determined from the cumulative distribution of noise levels, as shown in Figure 13(b). The most used statistical centile sound levels are L_1, L_{10}, L_{50}, L_{90}, L_{95} and L_{99}, shown in the example in Figure 13. Particular levels are L_0 and L_{100} corresponding to the maximum L_{max} and minimum L_{min} values of the sound level respectively.

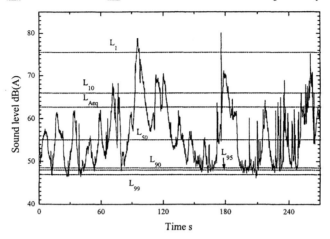

Figure 13(a): Statistical centile levels and cumulative distribution of noise levels.

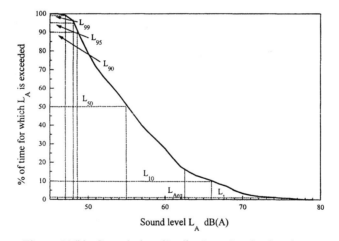

Figure 13(b): Cumulative distribution of noise levels.

Levels exceeded for small values of n, such as L_1 and L_{10}, are used to represent the more intense short-duration noise events. In Australia and the United Kingdom L_{10}, determined on the 18 hour period from 6 to 24 hours, is used as the target value for new roads and for insulation regulations. Levels exceeded for large values of n, such as L_{90}, L_{95} and L_{99}, are used to represent the residual level, little influenced by discrete noise events. The noise climate is usually defined as the difference $L_{10} - L_{90}$ and gives a satisfactory indication on the time variability of the noise levels.

2.4   Equivalent continuous sound level

Even if statistical centile levels enable us to summarize a time-varying noise in a set of a few values, such as L_{10} and L_{90}, there is still the need, especially for regulation purposes, to describe such a variable phenomenon by a single numerical value. This can be accomplished by using the equivalent continuous sound level L_{eq}, usually expressed in dB(A), defined as the level of a steady continuous sound containing the same sound energy over a defined measurement time T as the actual time-varying sound level $L_A(t)$, that is:

$$L_{Aeq} = 10 \cdot \log_{10}\left[\frac{1}{T}\int_0^T 10^{L_A(t)/10}\,dt\right] \qquad \text{dB(A)} \qquad (29)$$

Eqn (29) is actually a logarithmic form of eqn (2), giving the linear rms value of the sound pressure over a time period T. The L_{Aeq} level may be measured directly by an integrating-averaging sound level meter (see 3.2)[16]. Since the L_{Aeq} level is a measure of exposure which accounts both for magnitude and duration of the noise, it has become one of the most widely used "scales" for evaluating environmental noise and the resulting annoyance. However, as a measure of noise annoyance the equal energy principle at the basis of L_{Aeq} appears to be inadequate in some circumstances, for instance when occasional high-level noise events of short time duration occur. In fact the energy of these events is spread into the quieter parts by the time averaging process, as shown in Figure 14, and the resulting L_{Aeq} may underestimate the annoyance due to such events.

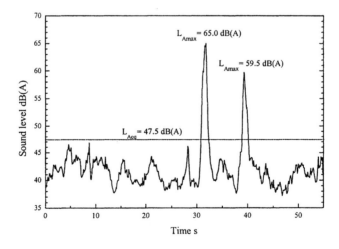

Figure 14: Noise events and L_{Aeq}.

The measurement time T ranges from seconds to hours, the latter being limited by the power supply available for hand-held instruments (commonly provided by batteries). Most common values are 1 hour or longer periods, such as day-time (usually from 06.00 to 22.00 hours), night-time (frequently from 22.00 to 06.00 hours) and 24 hours.

For free-flowing road traffic the following empirical relationship between L_{10} and L_{Aeq} has been found[17]:

$$L_{10} \cong L_{Aeq} + 3 \qquad\qquad \text{dB(A)} \qquad (30)$$

However, eqn (30) is not valid for vehicle flows of less than about 100 vehicles per hour.

The use of the L_{Aeq} level is widespread all over the world and forms the basis for rating community noise in several countries; it was recently proposed as the descriptor of transportation noise to be used in the member states of the European Union.

2.5 Single event noise exposure level

This scale, usually denoted as SEL or L_{AX}, is defined as the level of a continuous sound lasting 1 s which contains the same sound energy as the actual time-varying level $L(t)$ of the noise event and is given by:

$$SEL = 10 \cdot \log_{10} \left[\frac{1}{T_{\text{ref}}} \cdot \int_{-\infty}^{\infty} 10^{L(t)/10} \, dt \right] \qquad\qquad \text{dB} \qquad (31)$$

where T_{ref} is the reference time equal to 1 s. In practice the integration is limited to the time during which the actual noise level is within 10 dB of its maximum value $L_{\max}$ as follows:

$$SEL = 10 \cdot \log_{10} \left[\frac{1}{T_{\text{ref}}} \cdot \int_{t_1}^{t_2} 10^{L(t)/10} \, dt \right] \qquad\qquad \text{dB} \qquad (32)$$

where t_1 and t_2 denote the beginning and end of the noise event respectively, as shown in Figure 15. The time interval $t_2 - t_1$ is usually referred to as conventional duration of the noise event. SEL is basically an equivalent continuous level L_{eq} normalized to a time period of 1 s and can be derived from L_{eq} as follows:

$$SEL = L_{eq} + 10 \cdot \log_{10} \left(\frac{T}{T_{\text{ref}}} \right) \qquad\qquad \text{dB} \qquad (33)$$

where T in s is the measurement time of L_{eq}. For most transient sounds, such as an aircraft or a vehicle passing by, the A-weighted sound level $L_A(t)$ is usually used instead of the unweighted level $L(t)$, while the C-weighted sound level $L_C(t)$ is preferred for short-duration, high-energy impulsive sounds.

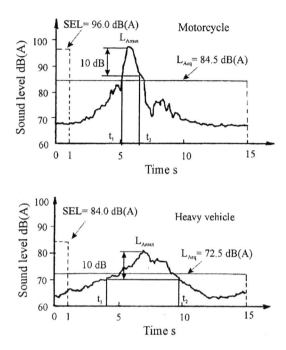

Figure 15: Definition of the single event noise exposure *SEL*.

The usefulness of *SEL* is evident when a number of different types of noise event occur. These may differ because of the operating conditions or individual characteristics of the same type of source, such as aircraft, or the occurrence of two or more totally different types of noise source, for example aircraft and train. In either case the knowledge of SEL_i for each type of event, further categorized in terms of operating conditions where applicable, enables us to calculate the equivalent continuous level L_{eq} over the period T by the following relationship:

$$L_{eq} = 10 \cdot \log_{10} \left[\frac{1}{T} \cdot \sum_{i=1}^{n} 10^{SEL_i / 10} \right] \qquad \text{dB} \qquad (34)$$

where n is the number of events in the time T.

2.6 Day/night equivalent sound level

This rating, denoted as L_{dn}, is based on L_{Aeq} and was proposed for community noise assessment by the US Environmental Protection Agency[18]. The L_{Aeq} is determined over 24 hours but the noise levels during the night-time period, defined from 22.00 to 07.00 hours, are penalized by the addition of 10 dB to take into account the increased annoyance caused by noise in this period. Thus:

$$L_{dn} = 10 \cdot \log_{10}\left[\frac{1}{24} \cdot \left(15 \cdot 10^{(L_{Aeqd}/10)} + 9 \cdot 10^{(L_{Aeqn}+10)/10}\right)\right] \text{dB(A)} \quad (35)$$

where L_{Aeqd} is the equivalent continuous level during the 15 hour day-time period from 07.00 to 22.00 hours and L_{Aeqn} is the equivalent continuous level during the 9 hour night-time period. L_{dn} is not intended as a "single source" measure and it does not account for tonal components or impulse noise. In addition, rare loud events may not be adequately accounted for, and there is some question about the value of the night-time penalty. However, L_{dn} is widely used in the USA for rating all community noise and it is also applied in social surveys on people's reactions to community noise, as the sound exposure is frequently expressed in terms of L_{dn}.

In some countries a third time period is also considered, usually denoted as the evening-time period and defined within the day-time period, during which a penalty of 5 dB is applied to the noise level.

The definition of number and duration of the time periods into which the 24 hours are divided is a political decision depending on several factors, including the social habits and way of life of the population which, in turn, are influenced by cultural heritage and climate.

3 Measurement instrumentation

The objective of any noise measurement must first be clearly defined before beginning the measurement itself and the data necessary to achieve this objective should be specified. Following this stage, the choice of instrumentation has to be done also taking into account the practical aspects of the measurement which may lead to severe constraints.

The characteristics of the sound itself are important when choosing suitable instrumentation. The sound spectrum may be broad band, narrow band or highly tonal. The sound level may be fairly constant, highly time dependent or intermittent, with intense sound events separated by long quiet periods. Additional problems are encountered when measuring impulsive noise.

The type and deepness of analysis to perform on the data also influences the choice of both instrumentation and measurement procedures.

In all environmental noise problems, the equipment has to be used in the field and, therefore, it must be portable, that is, easy to set up and calibrate on site as well as being independent of external power supply and reasonably light in weight.

A wide variety of different systems, some consisting of a number of interconnected instruments, are available for the measurement of noise and cover most situations and needs to be found in practice. The huge and rapid progress of the digital technique has stimulated the manufacture of more and more sophisticated equipment which would otherwise become obsolete.

For this reason this section is limited to describing the basic components of a field measurement system, referring to the manufacture's literature for additional and updated information.

3.1 Microphones

A microphone is a transducer converting the sound pressure fluctuations into time-varying electric signals. It should meet the following requirements:

- linear frequency response over a wide frequency range;
- linear relation between output signal level of the microphone and sound level at the microphone over a wide range of sound pressures and at all frequencies within the useful range of the microphone;
- sensitivity not varying with time or with ambient conditions;
- small disturbance on the sound field due to the presence of the microphone.

The condenser microphone is best able to meet all the above conditions and has therefore become the most widely used. It is based on the principle that the capacitance of two electrically charged plates changes with their separation distance. One of these plates, the diaphragm, is extremely light and is moved by the sound pressure fluctuations; the resulting change of capacitance is fed to the instrument connected to the microphone. Because the capacitance is proportional to the plate area, the size of the microphone must be increased to achieve higher sensitivity. Unfortunately, this conflicts with the requirements for a wide range of linear frequency response and omnidirectivity, both of which improve if the diaphragm is small compared with the wavelength of sound being measured. Microphones having a diameter of 13 mm (½ inch) are preferred for most noise measurements because they are essentially omnidirectional (at frequencies below approximately 5000 Hz) and combine adequate sensitivity (around 12.5 mV/Pa) with a frequency range covering the audible sounds. A direct-current (DC) voltage must be applied to the backplate and the diaphragm and this makes the microphone susceptible to high humidity since electrical leakage can occur. The polarization voltage, of the order of 150 to 200 V, is supplied by a preamplifier which also has the correct impedance for connecting the microphone to the rest of the measurement system.

Electret-capacitor microphones operate by the same principle as condenser microphones but they do not need an external DC polarizing voltage, as the electric field between backplate and diaphragm is established by charges "permanently" trapped in or bound into a special polymer material such that a preponderance of positive charges resides on one side of the material and negative charges on the other.

When choosing and using any microphone, the type of sound field to be measured should be considered. In fact the frequency response of the microphone is influenced at high frequencies by the reflections and diffraction caused by its own presence in the sound field, and is therefore dependent to some extent on the direction of the incident sound. As a consequence it is necessary that the characteristics and orientation of the microphone chosen are suitable for the type of sound field being investigated. Microphone characteristics are usually reported as free-field, pressure or random incidence response. A free-field microphone is designed to compensate for the disturbance caused by its own presence in the sound field, provided that the direction of propagation of the sound wave is perpendicular to the microphone's diaphragm (frontal incidence, 0°); therefore such a microphone should be pointed toward the noise source, as shown in Figure 16. A pressure microphone has a linear frequency response to the sound field as it exists, including its own disturbance, and should be used for measurements in small volumes, such as an artificial ear; this microphone should be held at right angles (grazing incidence, 90°) to the direction of sound propagation, as shown in Figure 16. A random incidence microphone responds uniformly to sound waves arriving simultaneously from any direction and, therefore, it should always be chosen when investigating diffuse fields. When used in a free field, a random incidence microphone should be pointed at an angle of 70° to 80° to the source, as shown in Figure 16.

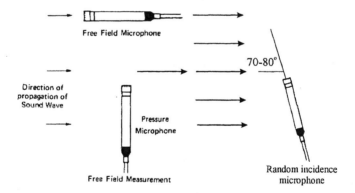

Figure 16: Microphone orientation in the sound field.

Outdoor noise measurements are very often taken in the presence of wind. In a moving airstream any microphone produces turbulence which deflects the diaphragm and, therefore, generates a spurious signal interpreted as sound by the measurement system. To reduce such a spurious signal fitted windscreens are recommended on the microphone. They are made of light foam material essentially transparent to acoustic signals at frequencies up to 3 kHz. The energy in a spurious wind induced signal is greatest in the low frequency range of the spectrum; thus, A-weighted sound levels are not as significantly affected by wind noise as unweighted sound levels.

3.2 Sound level meters

Among the different types of instruments available to measure sound levels, the most widely used is the sound level meter which enables us to measure the frequency-weighted and time-averaged sound pressure level. Sound level meters have been improved throughout the years and now are hand-held instruments, light in weight and battery-powered.

The principal components of a sound level meter, shown in the block diagram in Figure 17, include the microphone, a preamplifier, a frequency weighting network, an amplifier, a level range control (not always present), a time averager and an indicating device.

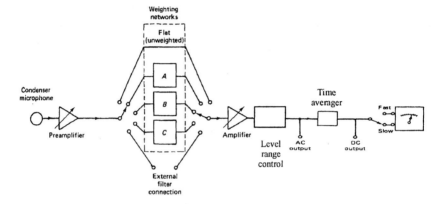

Figure 17: Block diagram of a typical sound level meter.

The electrical signal from the microphone is fed through the preamplifier, usually contained in the same housing as the microphone, to the frequency weighting network. This section enables us to select the A, B (not available in all sound level meters), C weighting or to leave the signal unweighted (linear). The last option is normally used when the spectrum of the noise is of interest; in some sound level meters it is possible to connect the instrument to an external set of frequency filters, usually octave and/or 1/3 octave band filters described in 1.2.

After further amplification, the level range control enables us to adjust the range of sound levels that can be measured for a given setting of the controls. The adjustments usually are in 10 or 20 dB steps. Digital sound level meters, having a nominal operating range of 90 dB or more, do not require such a control.

The time averager squares the signal, averages it and takes the logarithm. Since the 1930s a running time average sound pressure with exponential time weighting, according to eqn (3), has been implemented in analog sound level meters. The "Fast (F)" reading of the signal is taken with an RC time constant of the averaging circuit equal to 125 ms, while the "Slow (S)" requires an RC value of 1000 ms. In some sound level meters an "Impulse (I)" reading is also available and it is used for measuring impulsive sounds; it has a time constant RC equal to 35 ms for sound pressure that increases with increasing time and 1500 ms for one that decreases with increasing time. Specifications for exponential time weighting are standardized internationally[19].

The advent of digital technology in the 1970s made it possible to provide time averaging without exponential time weighting, that is linear time averaging according to eqn (2). Integrating-averaging sound level meters operate in this way and their specifications are standardized internationally[16].

Two outputs are frequently available, providing an alternating (AC) or direct (DC) current signal to be used as an input to other instrumentations, such as tape recorders, frequency analyzers (including FFT analyzers) and graphic levels recorders.

The indicating device displays the numerical value of sound level in dB and this function may be fulfilled in an analog, quasi-analog or digital way. An analog readout is given by a pointer moving across a graduated scale calibrated in dB. This readout may be simulated by a quasi-analog indicator in the form of a bar graph, usually available on liquid crystal displays (LCD). A digital indicator displays the numerical value of the sound level meter, frequently with one decimal digit, and provide additional information (overload, display mode, battery condition and so on).

Sound level meters should comply with the relevant national or international standards[16], [19]. These standards classify sound level meters into four categories depending on their accuracy as follows[19]:

- Class 0, prescribing the most stringent tolerances (±0.4 dB) with respect to linearity of level, deviations in frequency response and from omnidirectionality, to be used for laboratory reference purposes, when extreme precision is required;
- Class 1, defining a precision sound level meter having an engineering grade accuracy (±0.7 dB) which must be satisfied over a wide range of temperature and humidity required for field measurements;
- Class 2, corresponding to a general purpose sound level meter complying with no stringent tolerance (±1.0 dB) and having only A weighting, with the others optional;
- Class 3, corresponding to the simplest type of sound level meter with the least stringent tolerance (±1.5 dB), to be used for survey measurements.

The most recent models of sound level meter, taking advantage of digital techniques, are able to perform many calculations and functions, such as simultaneous Fast, Slow and Impulse detectors, determination of the statistical centile levels, cumulative and distributive analysis of sound levels, storing the time history of sound levels for periods defined by the user, conditioning storage of noise events based on exceeding a user-defined threshold level for a finite time period, real time frequency analysis in octave or 1/3 octave bands, and so on. Great changes in the structure of the instrument have also occurred and nowadays a notebook computer can perform more functions than a conventional sound level meter, providing it is connected to a data acquisition unit equipped with a microphone transferring the data in real time.

3.3 Calibrators

Prior to any set of measurements it is highly recommended, if not made mandatory by legislation, to check the accuracy of the sound level meter. This calibration provides for the consistency in measurements, allows more accurate comparison of measurements made over long time periods, and points out any change in the accuracy of the instrumentation. The calibration has to be performed before and after any set of measurements and the difference in the corresponding sound levels should be within a specific range. The operation is also necessary when a tape recorder is used; the known sound level at a fixed frequency emitted by the calibrator must be recorded on the tape to make it available as a reference from which to calibrate any instruments which may be used in subsequent analysis.

The calibration must not be confused with the check of conformity to the specifications relevant to the class of accuracy of the sound level meter. This check has to be performed periodically in the laboratory and requires more accurate techniques and sophisticated equipment as it is aimed at verifying the amplitude and frequency response of the sound level meter.

Two battery driven portable acoustic calibrators, working on slightly different principles, are available to accomplish the calibration conveniently under field conditions.

A pistonphone is an accurate, reliable device which, by means of moving pistons, produces a highly stable and distortion free sinusoidal sound pressure variation in the pistonphone's cavity which is closed on the opposite side by the microphone. A typical pistonphone signal is 124 (± 0.15) dB at 250 ($\pm 1\%$) Hz. The second calibrator is based on a tiny loudspeaker, driven by a stabilized electronic oscillator, producing the nominal sound pressure level in a small cavity into which the microphone is inserted. A typical acoustic calibrator signal is 94 (± 0.3) dB at 1000 ($\pm 2\%$) Hz. Calibrators operating at several frequencies and sound levels are also available, enabling calibration near sound levels and frequencies to be measured.

3.4 Other equipment

It is often more convenient and sometimes necessary to record and store the sound signal, so that it can be reproduced later for analysis. Among the many reasons for doing this there are the following:

– reduction of the time spent in the field and the equipment used there;
– analysis on the same data by different techniques which could not necessarily be carried out in the field;
– non repeatability and/or uncertain occurrence of certain types of noise event (sonic booms, explosions).

For such situations the tape recorder is the most universally used instrument. For use in the field it should be portable, battery operated, reliable, easy to operate, having a wide dynamic range, low wow and flutter and a flat response over the frequency range of interest. Tape recorders based on the digital audio tape (DAT) technique comply with most of the above requirements.

The spectrum of stationary sounds can be determined using a set of filters, usually octave or 1/3 octave bands, swept in sequence at discrete steps over the frequency range of interest. This procedure is not adequate for time varying sounds which require real-time analysis using parallel filter banks. To perform such analysis real-time frequency analyzers are available and portable models have been manufactured to enable measurements in the field.

4 Noise surveys

Throughout the years a large number of surveys aimed at describing the acoustic field in a community have been conducted. Most of these surveys deal with the measurement of the outdoor noise environment in inhabited areas. One of the earliest example of such surveys is that carried out by the New York Noise Abatement Commission[20] in 1930, established to "study noise in New York City and to develop means of abating it". The sources of noise that were identified as causing problems are shown in Figure 18. Most of these noises are still topical today, and it is perhaps an indictment upon society that their mitigation is confounded by both political and economic aims. Even if many reports are available in the literature, they are not the totality of surveys undertaken, as those conducted by city and state authorities often might not be reported. These studies have variously been termed; most frequently they are reported as community noise (referring to outdoor noise in the vicinity of inhabited areas) or ambient noise (the all-encompassing noise associated with a given community site).

These noise surveys are time consuming and costly, as they require a large investment of resources in equipment and manpower. Therefore a great deal of effort should be put into trying to achieve the best compromise between the amount of resources expended and the reliability of the collected data in order to increase the cost efficiency. The property of the surveys, to be specific for both the site and the aim of the study, makes it very difficult or even impossible to compare the data collected in different studies, for example the acoustic environment of one city, or the noise exposure of its residents, with those of another. To overcome this difficulty, the TNO[21] have recently started a project, aimed at creating an archive where data from different surveys are collected and made comparable, in order to provide the basis for drawing further inference from the statistical analysis of the pooled data.

As any noise survey has to be tailored to local physical conditions and to specific objectives, designing the ideal survey is not feasible; on the other hand it would be practicable to provide guidelines on which any future survey can be designed to be both appropriate and effective.

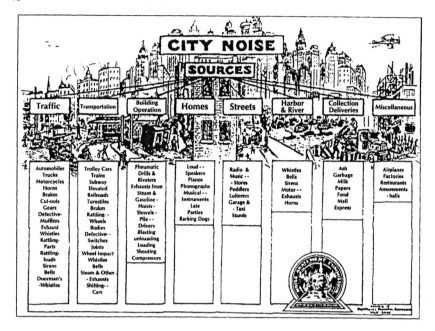

Figure 18: City noise sources[20].

4.1 Purposes of noise surveys

Given the wide range of purposes for which the noise surveys are made, they vary widely in depth and detail. In addition, they rarely have only a single objective and the multiple objectives might lead to confusion and compromise in the selection of sampling strategies. The purpose of a noise survey heavily influences the type and depth of the measurements to be undertaken. Typical purposes are the following, most of them already catalogued by Bishop and Schomer[22]:

1) To compare sound levels with limits specified in noise legislation.
2) To determine the suitability of land for different uses and activities, i.e. involving the comparison of existing or future noise environment with land use criteria and noise zoning.
3) To obtain environmental descriptions for assessing current or future noise impacts.
4) To determine the needs and/or extent of noise control actions on existing or future noise sources.
5) To identify outdoor noise sources and determine their contribution to the ambient noise.
6) To obtain a description of community noise for correlation with the community's response to noise.
7) To estimate the noise exposure of individuals.
8) To support the formulation of legislative and planning actions to reduce community noise exposure.

4.2 Characteristics of urban noise

A general description of the acoustic field in an urban area has been given by Brown and Lam[23]: it is composed of innumerable different types of noise sources, some intermittent, some continuous, some local, some distant, some high level, some low level, some broadband, some narrowband, some impulsive, some steady state, and so on. It is not surprising that the noise sources listed in Figure 18 are still of great importance in today's urban environment, as they are strictly related to human activities. The importance of each source depends on the site and the surrounding area.

Normally the predominant source in the majority of sites is road traffic noise which, indeed, is nowadays the sound-track of everyday life. Several sound sources related to vehicles (klaxons, alarm and emergency vehicle sirens, air brakes and so on) can become important especially at night. The increase of railway traffic is also a source of concern, as well as the rise of aircraft operations in airports which, due to the expansion of urbanization, are also getting closer and closer to residential areas.

In addition to transportation noise other sources need to be considered, especially at night and/or in specific seasons. Most of them are related to recreational activities, such as restaurants, pubs, discotheques, pop concerts, sporting events, fun fairs, fireworks and so on. Even if some of these sources occur occasionally, their noise immission can be high enough to produce a significant impact on the exposed population.

Noise from construction sites and maintenance road works are also important sources of noise, the former being present not only in suburban areas but also downtown where old houses are demolished to rebuild new and taller buildings, and the latter being carried out at night to reduce the interference with road traffic. For the same reason nowadays garbage collection is also frequently executed at night, so becoming one more source of noise annoyance.

Neighborhood noise, formed by a wide variety of sources including people and animals, is another widespread source especially in high density population areas.

For most cities the basic pattern of noise initially results from a network of road line sources, with the line sources having a hierarchy of intensities from the busiest streets (top) to the local roads (bottom), the latter more correctly being described as moving point sources. Noise from this network decreases with increasing distance from roads by hemicylindrical spreading, modified by the presence of buildings and ground effect in the urban configuration. Superimposed on this pattern are the point noise sources, such as industrial plants, air conditioners, etc., again with a hierarchy of intensities but most with a range of influence smaller than the noise from the road network. Noise spreads hemispherically from these sources. Also superimposed is the noise from aircraft and railway traffic: any single aircraft and train produces large variations in the acoustic field over space and time.

As already said, in most of the cities the noise from road traffic dominates the largest part of the urban area, except at locations near to point sources and when aircraft fly overhead or trains pass nearby. In areas where the spacings between the road line sources are large, noises from the point sources begin to dominate. At night road traffic flows drop on all roads, effectively increasing the spacing between line sources, and again point sources become relatively more significant.

The superpositioning of noise from all the above sources results in a very complex acoustic field, which is characterized by large variations in noise levels as one moves from the influence of one source to another and also by large variations of noise levels over time. Much of the planning effort in noise surveys is concerned with the development of methods for coping with such spatial and temporal variations in sound levels.

4.2.1   Spectral characteristics

The spectral content of community noise usually shows very large variations due to the wide variety of noise sources concurring to it. However, when road traffic is the predominant source, the 1/3 octave-band noise spectra generally show the pattern reported in Figure 19. The 1/3 octave-band levels are irregular in the low frequency range up to 125 Hz and decrease with frequency at rates of 3 to 6 dB/octave at higher frequencies. Most of the energy is contained at low frequencies, due to the pervasiveness of road traffic and other mechanical noise sources present in the urban areas. Many local or intermittent noise sources can produce spectra different from those shown in Figure 19, particularly at frequencies above 1000 Hz, but they tend to those shapes with increasing the distance between source and receiver, the low frequencies being less attenuated than the high frequencies.

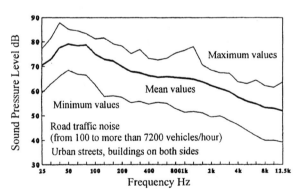

Figure 19: 1/3 octave-band spectra of road traffic noise in urban areas[24].

Spectral analysis is required in detailed noise control studies to optimize the acoustical efficiency of the actions to implement. Octave-band or 1/3 octave-band spectral analysis is adequate in most of the circumstances. For example, the reference spectrum of road traffic noise, as reported in the EN 1793-3 standard[25] issued by CEN (European Committee for Standardization), is given in Figure 20. The spectrum, normalized to the 1/3 octave-band level at 1000 Hz, should be applied in test methods for determining the acoustic performance of traffic noise reducing devices, such as barriers.

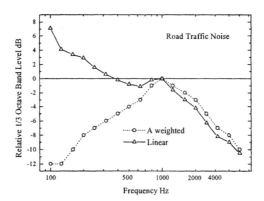

Figure 20: Reference 1/3 octave-band spectrum of road traffic noise[25].

4.2.2 Temporal patterns of sound levels

The temporal pattern of sound levels at a given position depends on the type of the major sound sources forming the ambient noise. Notwithstanding the large variability of such a pattern, normally a fairly steady low sound level is identifiable on which the sound levels due to discrete single events are superimposed (for example vehicles and/or train pass-by and/or aircraft flyover). Figure 21 shows the sound level time history in a site where the ambient noise includes contributions from distant sources and local sources (car and aircraft) which produce discrete noise events.

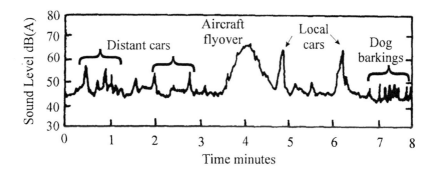

Figure 21: Example of a time history of sound level.

To obtain concise and meaningful descriptions of sound level time history statistical centile levels, described in 2.3, are frequently used.

Sites exposed to moderate and high flows of road traffic, with no other important sources, generally show a near-gaussian distribution of sound levels, as shown in Figure 22(a)[26].

Under other conditions often encountered in urban areas, such as stop-and-go traffic flow due to traffic lights, slow-moving or low traffic flow, differences from the gaussian distribution occur, as shown in Figure 22(b)[26]; these differences may be measured by skewness and kurtosis of the sound level distribution.

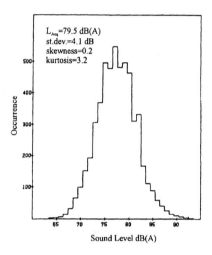

Figure 22(a): Sound level distribution in a high traffic urban street[26].

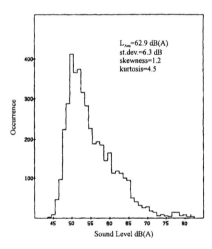

Figure 22(b): Sound level distribution in a low traffic urban street[26].

Figures 22(a) and 22(b), both corresponding to the distribution of sound level measured every 0.1 s for 1 hour continuously, show that at low traffic flow (occurring frequently at night) the distribution has bigger values of skewness and standard deviation.

The hourly values of the most used statistical centile levels for describing the road traffic noise in urban areas can be predicted from the hourly L_{Aeq} level as follows[27]:

$$L_1 = 0.98L_{Aeq} + 10.8 \quad \text{dB(A)} \qquad L_{10} = 1.05L_{Aeq} - 1.1 \quad \text{dB(A)}$$

$$(36)$$

$$L_{50} = 1.08L_{Aeq} - 10.2 \quad \text{dB(A)} \qquad L_{90} = 0.99L_{Aeq} - 9.7 \quad \text{dB(A)}$$

Figure 23 reports the L_{Aeq}, L_{10}, L_{50} and L_{90} values, averaged over the indicated number of sites divided into three different traffic flows (low, moderate and heavy). The data refer to measurements taken at the street kerb for a 10 minute period from 09.00 to 12.00 hours during weekday mornings.

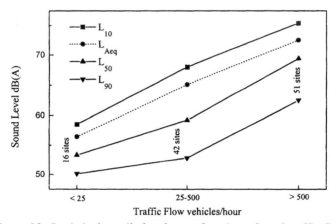

Figure 23: Statistical centile levels as a function of road traffic flow.

As already pointed out, noise levels due to urban road traffic typically vary with time within the day, from day to day and, for holiday resorts, from season to season. Nevertheless they show similarities which might be useful for classifying such time-varying phenomena into categories, each of these being represented by a typical pattern. These patterns, based on a user-defined descriptor with a fixed time resolution (most frequently the hourly L_{Aeq} levels), can be fitted on the relevant data measured by temporal sampling techniques, so enabling us to predict the L_{Aeq} for the time period of interest[28]. For example, studies carried out in Italy and Spain have led to the time patterns of the hourly L_{Aeq} during the 24 hour period shown in Figures 24 and 25 respectively[29]. The patterns are in terms of the difference between the hourly L_{Aeq} and the 24 hour L_{Aeq}. As would be expected, the largest differences amongst the patterns occur in the night-time period.

The comparison of the time patterns obtained for the two countries indicates that the Spanish type 1 (high traffic flow) shows a similar shape to the Italian type 4 (medium traffic flow at weekend) in the period 1-19 hours and a time lag of one hour occurs from 13 h, while the Spanish type 2 (medium traffic flow) is fairly similar in shape to the Italian type 3 (medium traffic flow during weekdays).

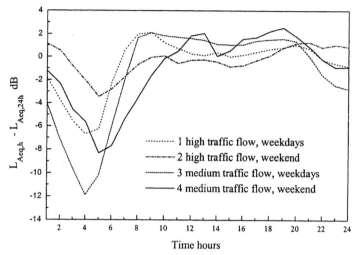

Figure 24: Hourly L_{Aeq} time history patterns for Italian cities[29].

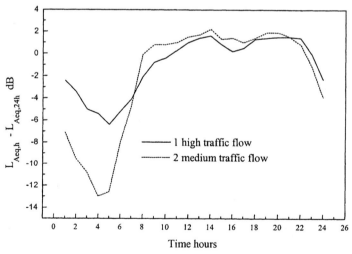

Figure 25: Hourly L_{Aeq} time history patterns for Spanish cities[29].

4.3   Measurement techniques

Safeer[30] clearly showed that community noise levels, varying as a function of a large number of variables, should be treated as a statistical phenomenon. As previously shown, the noise level at any point in a community is the result of a complex interaction of a large number of independent noise sources under varying atmospheric conditions and physical attributes of the area surrounding the measurement site, which may serve to reinforce or reduce the sound level between the source and the receiver. Thus, a single noise level value for a given moment in time at some fixed point in the community is not a reliable measure of the noise level in that community. In many cases it is not even a measure of the noise level at that particular point. The single noise level value, due to its variations, must be treated statistically and it may or may not, depending upon the degree of variation, be a good representation of the noise level at the measurement site. Similarly, depending upon the variation among sites, the single number for a single site may or may not be representative of the noise levels for a larger area. In addition, even if the single number at a single point can be considered representative of the noise level for some given points in time, it may or may not be representative of some longer time period.

In order to obtain a statistically reliable descriptor of the noise environment which is representative of a given area, the following three aspects must be considered:

- spatial sampling, namely number and location of the measurement sites in the selected area;
- temporal sampling, that is frequency and duration of measurements in each site;
- measurement setup, namely position of the microphone, environmental conditions and so on.

Often there may be an interaction between spatial and temporal sampling.

4.3.1   Spatial sampling[23]
In spite of the wide diversity of the noise surveys, they can be grouped into four categories depending on the spatial sampling technique as follows:

- random sampling;
- stratified sampling;
- receiver-oriented sampling;
- source-oriented sampling.

4.3.1.1   Random sampling   In this type of sampling the measurement sites should be arbitrarily located, but most often they are obtained systematically by a grid (usually rectilinear) superimposed on the map of the area to be surveyed. The size and boundaries of the grid mesh can generally be determined by an analysis of the land use and any identifiable noise sources. The dimension of mesh usually varies from 100 m to 1000 m and it would appear to be determined more by available resources than by notions of the spatial variability of the acoustic field. The mesh size should be smaller close to the source and larger going away from the source. The measurement sites are usually located at the grid intersections or at the center of each mesh of the grid.

Bias may be introduced by the need to displace the measurement site away from the original location when it falls in an inaccessible position (i.e. inside a building, or in the middle of a road).

Surveys based on this sampling are useful in providing exploratory information about the sources of noise in the community, and the shape and range of the temporal noise level distributions likely to be encountered.

One result which could be reported from these surveys is a cumulative distribution showing the proportion of measurement sites p_m at which predetermined noise levels were exceeded. If the sample is selected without bias and the sample size is adequate, the cumulative distribution can be interpreted in terms of the proportion of the study area p_a, not just the p_m. In this way random surveys can be quite successful in assessing the population of noise levels in an urban area.

However, random sampling is not an efficient way to gather urban noise data, where the predominant source is road traffic. In fact the distribution of noise levels throughout an urban area is highly positively skewed, as shown in Figure 26, being the long tail formed by the higher noise levels in the areas immediately adjacent to the roads and the bulk of the population consisting of the lower noise levels which principally occur away from the roads.

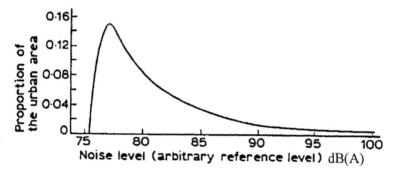

Figure 26: Simulated distribution of road traffic noise levels throughout an urban area[23].

Therefore, the sampling would be much more efficient if less effort were to be expended at sites with lower noise levels, and more effort expended at the higher noise level sites near the noise sources. This is the rationale of stratified sampling described in the next sub-section.

Information on the proportion of the area of a city which exceeds any particular noise level would allow inter-urban or even intra-urban comparisons of noisiness. However this information is of little use when the sound exposure of the inhabitants is of interest, unless it is known how the exceeded levels are distributed geographically relative to the receiver population.

4.3.1.2   Stratified sampling   The purpose of stratification in random sampling, providing that a fixed number of measurements are made within each stratum, is to reduce the possible variation in the statistical estimates, so making the survey more cost effective.

The selection of the strata depends on their homogeneity with respect to noise sources, which can be defined by land use categories (such as residential, commercial, downtown, suburban and so on), population density or location of noise sources. The last selection criterion, based on the proximity to the dominant noise sources, is more effective than land use categories, unless these are related to the distribution of the noise sources. Measurement sites within strata are chosen randomly or arbitrarily at locations thought to be representative of the stratum. However, the arbitrary sampling may lead to a systematic overselection of noisier sites.

The number of measurement sites within a stratum is a function of the expected variation in the chosen noise descriptor (most frequently L_{Aeq}). Figure 27 provides a guide for selecting the minimum number of measurement sites required to determine an average sound level within 90% confidence limits. For example measurements at eight positions are required to achieve an accuracy for the average sound level of ±2 dB at the 90% confidence interval if the standard deviation of the measurements is equal to 3 dB.

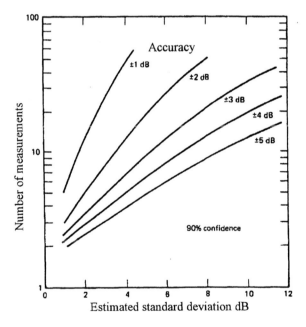

Figure   27:   Minimum   number   of   measurements   required   to   obtain   a predetermined accuracy.

4.3.1.3   Receiver-oriented sampling Surveys based on this type of sampling deal with the noise exposure of a particular class of receivers. Residential dwellings are likely to be the receivers of most interest, being where people spend a high proportion of their time, but also pedestrians or noise sensitive buildings (i.e. hospitals and schools) are often considered.

The sampling problem is quite different as we have to obtain a representative sample of the population of receivers rather than of the population of noise levels. Normally noise measurements are carried out at a fixed distance (usually 1 m) from the external façade of dwellings.

The major advantage of receiver-oriented sampling is that the results can be generalized directly to the target population if an adequate sample size is used. This sample size would be considerably smaller than required for sampling levels to produce a noise map of the study area and estimating the exposure of the target population from the map.

4.3.1.4   Source-oriented sampling In this sampling the noise measurements are made at locations selected either arbitrarily, for instance to represent different road and traffic conditions, or systematically, i.e. at equal distances along roads. It is important to point out that the results obtained by such a sampling technique with those given by the other procedures, such as random sampling, are not comparable.

4.3.2   Temporal sampling

In many circumstances it is necessary to determine the L_{Aeq} level over long time periods, for instance to check the compliance with noise limits issued by legislation. It has already been pointed out (see 1.4.6) that these long-term values are more reliable in describing community noise as they include the variety of weather conditions (particularly speed and direction of wind) which can have an effect on sound levels, increasing with the distance between source and receiver.

Nowadays powerful and advanced instruments are available for measuring noise over longer terms, such as one month or a year. They can be installed permanently at fixed positions, as are the noise monitoring units located around major airports to check compliance with the noise abatement procedures for aircraft operations, or easily removable after the measurement period in order to be installed in another site. However, these instruments are expensive and can not be used when the spatial sampling requires a large number of measurement sites, such as in an urban noise survey. Thus, there is frequently a need to compromise between the requirements of accuracy in determining the long-term value of L_{Aeq}, achievable by continuous monitoring, and the restrictions due to the limited resources of the equipment, personnel and time.

This problem can be satisfactorily managed using temporal sampling, by which the long-term value of L_{Aeq} is estimated from the L_{Aeq} values of noise samples taken at more or less regular shorter periods identified within the long-term T_L. These periods can be defined as follows:

- reference time T_R, determined within the 24 hour period; usually two reference times are distinguished, namely day-time (generally from 06.00 to 22.00 hours) and night-time (most often from 22.00 to 06.00 hours);
- observation time T_O, determined within the reference time which can include one or more T_O, not necessarily having the same duration; each T_O has to be selected in order to contain the noise having the characteristics of interest;
- measurement time T_M, determined within each observation time which can include one or more T_M, not necessarily having the same duration; each T_M has to be selected taking into account the time variability of the noise.

Considering the road traffic noise in urban areas, the long-term T_L usually includes either the weekdays (from Monday to Friday) or the weekend (from Saturday to Sunday). The observation time T_O is frequently taken equal to one hour. Figure 28 shows an example of the above time periods for an urban street.

Several different schemes for temporal sampling of noise levels have been developed and they can be grouped into the two following categories[30]:

- continuous sampling;
- time compression sampling, sometimes called microsampling.

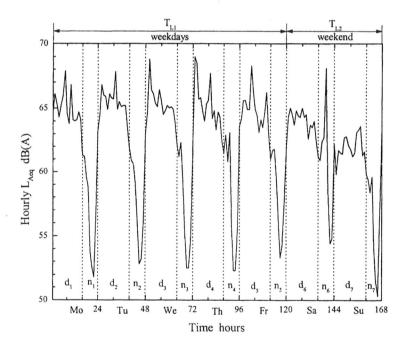

Figure 28: Time history of hourly L_{Aeq} in an urban street.

The continuous sampling involves collection of a continuous sample of noise for a duration of x minutes each hour during the reference time, where x minutes is less than 60 minutes. This scheme assumes that the statistical distribution of sound levels, obtained from each x minute sample, is fairly representative of the distribution which would be obtained from a continuous sampling over the full 60 minutes. This assumption can be made only under a very limited set of conditions (sound levels relatively homogeneous over the 60 minute period and small variance), unless errors greater than ± 2 dB are acceptable.

Time compression sampling is achieved by constructing an x minute sample from a series of sub-samples of shorter duration. For example, the noise might be measured a total of 10 minutes (600 seconds) during an hour, with the acquisition of twenty 30-second sub-samples. The sub-samples may be either equally or randomly spaced over the total data collection period. The magnitude of the errors due to this sampling are generally lower than those produced by continuous sampling. As the sub-sample duration decreases, the accuracy in the estimate increases.

The difference between the estimated L_{Aeq} obtained by temporal samplings and the value measured by continuous monitoring depends on the time variability of the noise environment: when the latter is small, i.e. close to a busy road, the former is reduced. For most situations of urban road traffic noise, where there are likely to be a relatively large number of events (20 or more) occurring per hour, sampling of 10 minutes per hour provides reasonable accuracy; if practicable, the 10 minutes should be composed of several shorter sub-samples distributed throughout the hour. In Figure 29 the probability for the estimate of the L_{Aeq} over the 5 weekdays (from Monday to Saturday) and the 16 hour day-time period to be within ± 1 dB is plotted as a function of T_M/T_L; the experimental data, represented by the circles, refer to a standard deviation of the 10 minute L_{Aeq} levels in the range $2 \div 3$ dB. For example, to have a value of 60% for the above probability, about 3 samples of 10 minutes in the 16 hour day-time period are required at least.

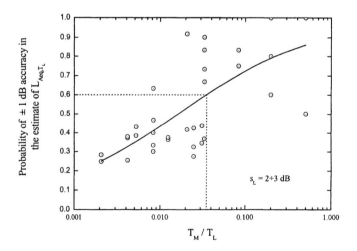

Figure 29: Probability of ± 1 dB accuracy in the estimate of $L_{Aeq,TL}$.

Potential for continuous monitoring over the reference time can be improved significantly in many situations by sampling noise at intervals at other auxiliary positions (slave sites) in the vicinity of the monitor location (master site). A comparison of the short sample levels with those measured at the master location at the same time will establish the differences in the noise environment at the slave positions with respect to the master site and will enable us to estimate the L_{Aeq} over the reference time at the slave positions from a limited sampling base. Similarly, long-term levels can be predicted accurately by a comparison of short-term (over some days) monitoring data obtained at one site with continuous (long-term) noise monitoring data at another site[22].

In temporal sampling it is important to collect data samples which are statistically independent in order to increase the statistical accuracy of measurements. Normally samples too close together in time, for example consecutive daily samples, shall not be inferred to be independent and they need to be tested for their statistical independence[31].

4.3.3   Measuring road traffic noise

In some countries the procedures for measuring road traffic noise are defined by either standards or legislation.

Normally, a free field microphone, protected by a windscreen and pointed towards the traffic lane, should be used. The microphone position, if not otherwise specified, should be at 1 m from the building façade exposed to road traffic and at a height of 4 m above the ground. This height, frequently corresponding to the first floor of buildings, would reduce spurious noise from people and improve safety for the instrumentation. To minimize the influence of the operator and observers the microphone should be mounted on a tripod and connected to the sound level meter with an extension cable.

Noises due to sources other than road traffic should not be included in the measurement either by interrupting the sound level meter during the measurement in the field or by eliminating them from the stored sound level time history later on. The presence of spurious noises, and the need for their suppression, is indeed a limitation for an extensive use of unattended instrumentation which, otherwise, would be appealing as it enables one to monitor noise over longer terms. Improvements have been made on this aspect in the recent instrumentation, such as conditioning audio recording of noise events based on exceeding a user-defined threshold level for a finite time period.

The wind speed should, preferably, be less than 2 m/s and measurements should only be made when the road surface is dry and without accidental irregularities (i.e. hollows).

The measurement at the building façade includes the sound reflection from the façade itself, which is estimated to increase the free field level of 2.5 dB. In highly populated metropolitan areas, the noise environment must be considered also as a function of height above the ground. Due to the reflections from adjacent buildings, the L_{90} value decreases slowly with height, while the lower statistical centile levels (such as L_1 or L_{10}), which are generally due to the stronger local sources, drop off more rapidly[22].

Most community noise surveys deal with outdoor noise. However, from the standpoint of defining the noise exposure of people during their daily life, outdoor measurements are inadequate and misleading because such data neglect the noise contributions of the many indoor noise sources and the noises arising from people's activities. Apart from these additional sources, the outdoor sound level is usually greater than the indoor one even when the window is open; in this case a mean value of 6 dB has been found for the difference between the outdoor sound level measured at 3 m from the building façade and the indoor one measured in the room at 1 m from the windowsill.

4.3.4   Measuring railway and aircraft noises

Procedures for measuring noise from these two sources, becoming more and more important in urban areas, are standardized[32], [33].

The main problem is to identify the noise events due to the source of interest (train pass-by or aircraft flyover) when other sources, frequently road traffic, are comparable in noise levels. To overcome this problem it could be necessary to select other measurement sites where either the masking noise is reduced or the source of interest becomes clearly predominant, as when getting closer to the source.

References

[1]   Ford, R.D., Physical assessment of transportation noise, *Transportation Noise. Reference Book*, ed. P. Nelson, Butterworths, London, pp. 2/3-2/25, 1987.

[2]   Beranek, L.L., *Noise and Vibration Control*, Institute of Noise Control Engineering, Washington, revised edition, 1988.

[3]   Hassall, J.R. & Zaveri, K., *Application of Brüel & Kjær Equipment to Acoustic Noise Measurements*, Brüel & Kjær, 3rd edition, 1978.

[4]   EN 60651/1994, *Sound Level Meters*, 1994.

[5]   Wahrmann, C.G. & Broch, J.T., On the averaging time of RMS measurements, *Brüel & Kjær Technical Review*, 2, pp. 3-21, 1975.

[6]   International Standard ISO 266-1975, *Acoustics - Preferred Frequencies for Measurements*, 1975.

[7]   Piercy, J.E. & Daigle, G.A., Sound propagation in the open air, *Handbook of Acoustical Measurements and Noise Control*, ed. C.M. Harris, McGraw-Hill, New York, pp. 3.1-3.26, 1991.

[8]   International Standard ISO 9613/2-1996, *Acoustics - Attenuation of Sound during Propagation Outdoors*, Part 2, A General Method of Calculation, 1996.

[9]   International Standard ISO 9613/1-1990, *Acoustics - Attenuation of Sound during Propagation Outdoors*, Part 1, Calculation of the Absorption of Sound by the Atmosphere, 1990.

[10]   Schultz, T.J., *Community Noise Rating*, Applied Science Publishers, London, 2nd edition, 1982.

[11]   International Standard ISO 226-1987, *Acoustics - Normal Equal-Loudness Level Contours*, 1987.

[12]   Stevens, S.S., Procedure for calculating loudness: Mark VI, *J. Acoust. Soc. Am.*, 33, pp. 1577-1585, 1961.

[13] Zwicker, E., Ein Verfahren zur Berechnung der Lautstärke, *Acustica*, **10**, pp. 304-308, 1960.

[14] ISO Recommendation R 532-1967, *Acoustics - Method for Calculating Loudness Level*, 1967.

[15] Stevens, S.S., Perceived level of noise by Mark VII and decibels (E), *J. Acoust. Soc. Am.*, **51**, pp. 575-599, 1972.

[16] European Standard EN 60804/1994, *Integrating-Averaging Sound Level Meters*, 1994.

[17] Alexandre, A., Barde, J-Ph. et al., *Road Traffic Noise*, Applied Science Publishers, London, 1975.

[18] U.S.E.P.A., Office of noise abatement and control, *Information on Levels of Environmental Noise Requisite to Protect Public Health and Welfare with an Adequate Margin of Safety*, Report N. 550/9-74-004, Washington, 1974.

[19] European Standard EN 60651/1994, *Sound Level Meters*, 1994.

[20] Noise Abatement Commission, *City Noise*, ed. E.F. Brown, E.B. Dennis Jr., J. Henry & G.E. Pendray, The Academy Press, New York, 1930.

[21] de Jong, R.G., Miedema, H.M.E. & Vos, H., Introduction to the TNO data archive, *Proc. of InterNoise '96*, Liverpool, vol. 5, pp. 2401-2405, 1996.

[22] Bishop, D.E. & Schomer, P.D., Community Noise Measurements, *Handbook of Acoustical Measurements and Noise Control*, ed. C.M. Harris, 3rd edition, McGraw-Hill, New York, 1991.

[23] Brown, A.L. & Lam, K.C., Urban Noise Surveys, *Applied Acoustics*, **20**, pp. 23-39, 1987.

[24] Pons, J. & Santiago, J.S., Traffic Noise Spectra in Madrid, *Proc. of 6th International FASE Congress*, Zürich, pp. 371-374, 1992.

[25] Standard EN 1793-3, *Road traffic noise reducing devices - Test method for determining the acoustic performance - Part 3: Normalised traffic noise spectrum*, CEN European Committee for Standardization, 1997.

[26] García, A. & Faus, L.J., Statistical analysis of noise levels in urban areas, *Applied Acoustics*, **34**, pp. 227-247, 1991.

[27] García, A. & Garrigues, J.V., 24 hour measurements of noise levels in urban areas, *Proc. of InterNoise '96*, Liverpool, vol. 4, pp. 2057-2060, 1996.

[28] Brambilla, G., Carletti, E. & Carretti, M.R., Assessment of L_{Aeq} level of urban noise by means of time history pattern, *Proc. of InterNoise '94*, Yokohama, vol. 3, pp. 1971-1974, 1994.

[29] Brambilla, G., García, A. & Garrigues, J.V., Classification of hourly L_{Aeq} patterns of urban noise, *Proc. of InterNoise '97*, Budapest, vol. II, pp. 903-906, 1997.

[30] Safeer, H.B., Community noise levels - a statistical phenomenon, *Journal of Sound and Vibration*, **26**, pp. 489-502, 1973.

[31] ANSI S12.9-1992, *Quantities and Procedures for Description and Measurement of Environmental Sound. Part 2: Measurement of long-term, wide-area sound*, 1992.

[32] International Standard ISO 3095-1975, *Acoustics - Measurement of Noise Emitted by Railbound Vehicles*, 1975.

[33] International Standard ISO 3891-1978, *Acoustics - Procedure for Describing Aircraft Noise Heard on the Ground*, 1978.

Chapter 3

Effects of noise on health

Michel Vallet
Inrets Case 24 69675 Bron Cedex, France

1 General introduction

Certain elements of our environment harm human health; noise is one of the nuisances we tend to think of first, though its impact on health is hard to identify. The effects of noise are fairly well recognized with respect to audition; however, they are underestimated for other health aspects. Recent research bringing to light non-auditive damage has not yet been diffused on a wide scale, although the nuisance caused by noise is now acknowledged.

1.1 Complaints against noise

Although complaints are not necessarily a good indicator of noise related nuisances, given their subjective dimension, their recent increase since 1990 has drawn attention from politicians, hitherto relatively insensitive to this environmental factor and its possible effects on health. Examination of the information on concern expressed throughout the population in random and quotas surveys highlights three trends:

For the vague formulation of a question such as "What are your reasons for concern about the environment?" noise does not stand out. In the survey made by Credoc in 1990, health problems linked to environmental pollution were mentioned by 24.6% of the French population, immediately after drinking water pollution. No spontaneous mention was made of noise.

In the same survey, a more precise question was asked: "What priority actions should be taken to combat environmental degradation?" The combat against noise appeared in third place (14.3% of the population) after the elimination of industrial waste (21.9%) and pollution of river and lake water (16.8%) but just before atmospheric pollution (13.4%).

In the Eurobarometer results of 1995, noise was considered as causing serious harm to the environment by 11 % of the European population, this percentage varying from 7% to 13% among member countries.

Regarding the reasons for complaining about the environment, the causes of nuisances, and daily stress, noise appeared to be the most frequently stated factor. Eurobarometer 1995: 31%, INSEE survey 1995 (France) 40%, in front of vandalism (33%), atmospheric pollution (18%), just after money problems (38%) and in front of unemployment (36%) and insecurity (26%).

In spite of the fragile nature of data provided by surveys, these figures demonstrate an interesting consensus on the subject of noise. Their magnitude has been analyzed in more scientific surveys on noise (cf. Chapter 4 by R. Guski). Furthermore, the importance of noise in our civilization has been confirmed by works seeking to estimate the number of persons subjected to different levels of noise. The figures obtained for certain countries are quite close, though less so for others. Lambert and Vallet[1] evaluated the total for road traffic:

- an average of 65% of the population in Europe is subjected to a noise exceeding 55 dB(A)
  in Leq 24h
- 17% are subjected to more than 65 dB(A)
- 1.4% above 75 dB(A).

According to the estimations of Stanners and Bourdeau[2], in countries such as Norway, Sweden and Finland, less than 20% are exposed to more than 55 dB(A). This implies that large portions of the populations of other countries are exposed to noise (Southern Europe).

If we now consider that several sources of noise can reach the same person (noise at work, noise during leisure, internal noise of transport) the percentage of persons exposed to more than 65 dBA **Leq 24h** has increased[3] from 15% in 1980 to 26% in 1990. This agrees with the results of investigations by surveys and demonstrates the magnitude of the problem.

1.2 Definition of health and wellbeing

Not all people exposed to noise will have health problems. However, it is worth recalling here the definition of health given by the World Health Organization: "Health is not only the absence of illness, but a complete state of physical, mental and social wellbeing". This definition from 1948 is highly suited to the effects of noise, which begin by a disturbance and a nuisance and can develop into more serious forms. In spite of the obvious generosity of the definition of health given by the WHO, in this chapter we have chosen to focus only on the severest effects of noise on man, while the more psychological aspects are dealt with by Rainer Guski in Chapter 4.

On considering all the works carried out over the last thirty-five years on the effects of noise on man, emphasis seems to be given to a notion of risk rather than that of causality, and the concept of sensitive groups. The definition of these sensitive groups is often approached in terms of socio-demographic factors: age, sex, level of education and income. Some works have attempted an approach in terms of overall sensitivity to noise during gestation, or according to sensorial hyperaesthesia, which is innate or acquired prematurely, or else according to the period of exposure. In spite of the inadequacy of the definitions given to groups sensitive to noise, these chapters will examine the greatest sensitivity and the greatest risk of appearance of noise effects on health for certain persons. This is because examining the causality of noise in the appearance of health problems is perhaps not as inextricable as a brief review might lead one to believe.

2 Mechanisms and types of noise health effects

Progress can be made in understanding the role of noise in the appearance of health problems by analyzing the noise effect mechanisms, which certainly vary according to the same type of effect examined.

2.1 Short and long term exposure to noise

In principle, it is easy to measure the amount of noise experienced by a person, characterizing his or her exposure at different times of the day, at home, at work, in different modes of transport, and during leisure, by wearing a noise dosimeter. In reality, we focus only on the source of noise in order to correlate the amount of noise and its biological response.

According to Rylander *et al.*[4], the dose-noise concept is used when observing different effects according to a noise level. When seeking the percentage of responses in a population, the term dose-response is used. When the noise intensity exposure increases, the type of response given by a person or a population can take different forms:

- an effect of adaptation for low doses (closing the window to listen to the television);
- repairable damage, which disappears when the noise stops (disturbed sleep);
- irreparable damage, observed after severe exposure to noise causing auditive problems (at work).

% of population

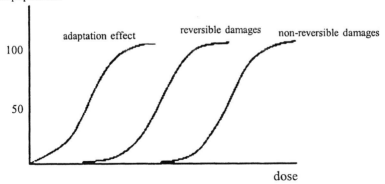

Figure 1: Noise-dose effects on the population (Rylander *et al.*[1]).

It can be seen clearly that for the same dose of noise, troubles can be simply adaptive or consist of repairable damage.

For environmental noise, the noise exposure time finally has the same importance as background noise in auditive problems. This time leads us, along with Laborit[5] "to abandon research into linear causality, the spirit of which has been passed down to us from the science of previous centuries, and which always seeks a cause for observed effects". This will certainly surprise acoustic engineers and decision-makers responsible for setting the doses and levels that should not be exceeded.

2.2 Habituation and adaptation

When raising the subject of noise in the environment, the persons questioned say they become accustomed to it. Physiologists consider that habituation corresponds to a disappearance of a reaction when a stimulus is repeated sufficiently. This habituation can be observed by the method of evoked cortical potentials, stimulated by very short bursts. The potentials disappear after a few dozen stimulations.

For longer exposures to more complex noises, no complete habituation of vegetative reactions have been recorded. As early as 1969, Jansen[6] demonstrated that, though not waking sleepers, noise levels of 55 dB(A) are associated with changes of stages of sleep during which a cardiovascular reaction occurs in the form of vasodilatation. Muzet[7] studied this vegetative aspect in greater detail by analyzing in the laboratory the vegetative reactivity of 26 subjects for 14 consecutive nights exposed to the isolated sounds of vehicles, varying from 40 to peaks of 65 dB(A), at a rate of 90 noises per hour. These studies highlighted a two-phase reaction of cardiac rhythm and vasomotricity in response to a peak noise from 60 to 65 dB(A). Cardiac acceleration and peripheral vasoconstriction occurred for each of these noises, followed, when the noise ceased, by sudden slowing down of the cardiac rhythm and vasodilatation.

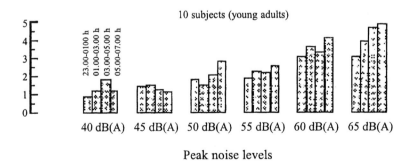

Figure 2: Comparison of the reaction to noise for a range of aircraft noise exposure.

During the same night, examination of the results by hourly periods shows that no habituation occurs. Furthermore, after 14 nights of exposure, whereas habituation exists on the subjective level and according to electroencephalographic (EEG) readings, this cardiovascular reactivity remains unchanged. These results, in particular that of the absence of habituation in the long term, were confirmed by Vallet[8] on persons exposed for five years to road noise. A short cardiac response persists through time, with an average magnitude of variation in the region of 10 beats per minute.

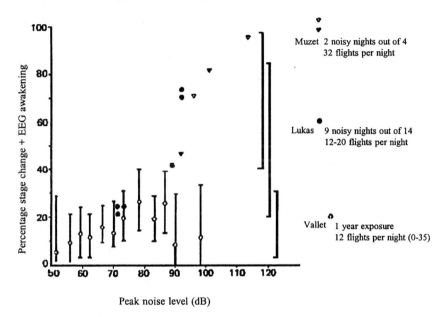

Figure 3: Comparison of laboratory (Muzet & Lukas) and field studies (% of people affected by noise).

The extension of the laboratory observation period and the in situ studies illustrated the conditions appertaining to the reaction rates and the noises causing awakening after long periods of nocturnal exposure.

The rates of waking caused by noisy events increases slightly with the noise made by airplanes flying at night. Habituation is not complete and, what is more, these results lead to the hypothesis of a reaction to the occurrence of noise according to an "all or nothing" mode, and that the frequency of "waking" responses to sound stimulation is not linked to maximum levels of sound, which is very closely correlated with the total energy of the passage of the airplane (SEL).

Adaptation occurs at a more global level and one of its aspects is voluntary effort. According to Laborit "action permits an organism to maintain its structure by converting the environment as well as possible for its survival". Although environmental noise is very frequent in industrial societies, it does not appear to threaten the survival of their populations. In developing countries, noise is a question of comfort and UNESCO considers that this concern should take as an example the sound emitted by a Rolls-Royce automobile, while the WHO states that "noise pollution is primarily a luxury problem for developed countries".

The view taken by Mouret and Vallet[9] seems more explicit. They think that "Survival of any species whatsoever depends on adaptation to the environment. Any adaptation to a given environment implies a strategy that can only be developed to the detriment of other functions, the choice of the best strategy in the short term being the prerequisite of any adaptation in the long term. However, our environment changes constantly, whether in terms of chemical composition or in terms of noise, whereas man most often only adapts by making use of external factors (air-conditioners, medicines, etc.). Naturally, we react to an uncomfortable and unexpected noise for several days. But its persistence through time causes us to react to its absence. So why worry, since we can adapt. Thousands of years of history have proven adaptation, as mentioned above, implies the loss of freedom and a cost. This cost is as physiological as it is psychological, insofar as these two facets of our organism can be separated. Is it not possible to consider deafness as an adaptation in the same way as callouses form on the hands of laborers? Adaptation, but at what level? Subjective? Objective? What then are the parameters to be studied, and for what length of time?

"Behavioral adaptation? Why not? Double glazing, sound proofing, moving house to a quieter neighborhood, but these objective signs demonstrate clearly that noise is a source of disturbance".

A good example of adaptation was observed by Rylander *et al.*[10] regarding sonic booms and their effect on sleep.

Adaptation is often limited and is rarely complete. For example, observations have shown that reactions during sleep to noisy events are ever present, even after several years' exposure to noise. Results of observations of subjects show that from 10 to 20% of airplane noise always lead to acceleration of cardiac rhythm. Sometimes apparent adaptation to a noisy situation can stop suddenly due to a breakdown considered as the final stage of stress (cf. below), despite there being no precursive sign of resistance to stress in the months before the breakdown.

2.3 Auditory and non auditory effects

Since 1993, several works have proposed inventories of various effects of noise on man (Schwarze and Thompson[11], Lindvall, Berglund[12], WHO[13], UK-IEH[14], Lercher *et al.*[15], Thompson[16], WHO[17], Muzet[18], Porter[19]). These summaries are often descriptive and report observations; they provide explanations of the variations of the frequency and seriousness of certain effects as a function of noise characteristics. Mouret's approach is more explanatory and medical.

"For a long time, and for some the period still exists, noise was considered as a physical phenomenon, having certain measurable parameters, acting via specific receivers, the ears, on a single system: the auditive system. We all know that this conception is wrong and that the effects of noise are not limited to audition any more than those of food are on the digestive system. The reactions it causes bring into play the entire organism, not only a juxtaposition of functional systems, causing it to react in a general way, by stress with its cardiovascular, neuroendocrinian, emotional, and attentional components, characteristic of the way in which most of our systems of defense and survival react.

"Besides its physical characteristics, noise has an informational and/or emotional content which, though difficult to quantify, exists nonetheless. As early as 1918, Hyde and Scalapino[20] showed that for the same sonic energy, changes in arterial tension differed among subjects when listening to a Tchaikovski symphony or the March of the Toreadors in Carmen. Thus it is clear that although objective as such, sound measurements will never be able to wholly explain the nuisance or effects of noise on health.

"The problem is all the more complicated as the possibility or impossibility of controlling sources of noise or stress, modifies or completely modifies a person's biological and psychological reactions. This therefore implies a fundamental difference, on the level of what is experienced, between the noise we generate and that which is imposed by others, even if they have the same characteristics".

The following is a brief recall of the auditive effects of noise. They have been described in detail with respect to noise at work for some time already (Ward[21]). The risks resulting from listening to music at loud volume were given in a bibliographical study by Loth *et al.*[22]: listening at 58 dB(A) for 100 minutes causes a TTS of 8 dB at 4 KHz. Exposure at a Leq of 89 dB (hard rock) and at 94 dB (classical music) from headphones for one hour causes a TTS of 4 or 5 dB at 4 KHz.

Thus, this temporary reduction can be recovered and is sometimes considered as auditive fatigue. Deafness can be observed in cases of exposure to intense noises for long periods, leading to a severe handicap for communication and a costly professional illness in countries where its treatment is reimbursed. Regarding the extra-auditive effects of noise, much work has been done in the area of cardiovascular response. Doctors have sought to highlight the chronic effects of noise as an extension of its brief effect on cardiac rhythm and arterial pressure, which are easy to illustrate. Highlighting a permanent effect caused by the temporary responses of the cardiovascular system in turn emphasizes that physiological effects that stop with stimulation have pathological effects that last even during periods of calm between two noisy situations and which are the symptoms of illness.

The major areas examined are:

- cardiovascular responses in children to environmental noises,
- cardiovascular effects in adults, whether these be responses to loud but brief noises or chronic effects. Regarding the latter, many epidemiological works have been carried out in industrial environments, but below we analyze, especially in section 4, works on environmental noise. The effects of noise on the cardiovascular system are the most visible of the body's vegetative

responses. Like the cardiovascular system, both the respiratory and digestive systems are not controlled by will. Noise causes respiration to accelerate which in turn leads to increased oxygen consumption. One of the most frequent problems of the digestive system is a reduction of salivation and intestinal action. Repeated modifications of the quantity of secreted gastric juices and their composition can lead to intestinal irritation, and possibly stomach and intestinal ulcers.

- effects on the endocrine system. These effects are less visible than those above, though they are the first to appear after audio stimulation, and they usually cause other effects. In brief, audio stimulation excites the hypothalamus by indirect, non-auditive nerve channels. The hypothalamus then triggers the release of ACTH (adrenocorticotropic hormone) by the pituitary gland. The latter, located at the base of the brain, produces ACTH which is transported by the blood and which in turn stimulates adrenocortical glands.

This chemical message incites the adrenocortical glands to produce anti-inflammatory corticoids (cortisol and cortisone), typical of stress. The effect of noise generates a neuro-hormonal transmission triggering – in the case of a loud or surprising noise – an excess of adrenaline in the medullo-adrenal and nerve endings. Stimulated by the noise, the brain and nerves exert their numerous actions via two other substances: norepinephrine and acetylcholine.

These modifications cause a considerable increase in the amounts of these substances when noises occur at night and this can be observed on examination of their excretion in urine. According to Muzet[18]: "The nocturnal increase of these rates can greatly affect the cardiovascular system, increase cardiac rhythm and arterial pressure, arrhythmia, platelet aggregation and also increase fat metabolism".

2.4 Noise and the immune system

The effects of noise on the immune system are partly linked to the modifications of hormonal secretions. During exposure to noise, considered as a stressful agent, the white blood cells undergo modification: according to Selye[23] the lymphocytes tend to disintegrate in the thymus, the lymphatic system and even in the blood itself. The eosinocytes tend to disappear in the blood. These cells regulate immunizing serological reactions as well as allergic responses and hyper-sensitivity to various foreign bodies. Logically, we agree with Muzet[18] that an organism subjected to repeated noise "is prone to weakening of its defenses, whether these be acquired or natural". Thus the organism becomes more fragile more globally and not merely for one specific system. Consequently, other factors such as physical and psychological stress combine with noise. Likewise, the hereditary or premature weakness of certain organs or important systems in the human body can increase the impact of noise.

Observations carried out on animals show that prolonged stress can, due to excessive secretions of glucocorticoids, lead to severe atrophy of the hippocampus (the part of the brain in close relation with the olfactory sensory system).

Other observations indicate that the concentration of magnesium in the cells is reduced, and that blood composition is modified. "A very extensive epidemiological survey carried out in the United Kingdom showed an increase in blood leucocyte concentrations in persons subjected to high levels of exposure to road traffic noise" (reported by Muzet[18]).

These different experimental results deserve clarification by more specific research in an area whose complexity is becoming increasingly apparent. This is the reasoning followed by Mouret and Vallet[9], after carrying out an experimental analysis of certain neurotransmitters playing a role in sleep.

"Overall, whether they are disturbed or not on the subjective level, persons suffering from chronic exposure to internal nocturnal noise levels in the region of 45 dB(A), show sleep structures similar to those of most depressive subjects and people whose age is more advanced than their own. These modifications can be reversed in a few weeks following sound-proofing of the bedroom or transfer of the latter to another part of the residence not exposed to nocturnal noise.

"However, what is most interesting in relation to the supposed role of serotonin is cardiovascular reactivity to noise during sleep. We have seen that during waking, noise is accompanied by an increase of cardiac rhythm, which after exposure of a few hours, recovers its normal rhythm in approximately 45 to 60 minutes. This reaction is both physiological and adapted, signifying the mobilization of the organism. Furthermore, habituation can occur during waking thus preventing metabolic overload.

"As seen further on, the problem is very different during sleep: the impact of noise on the cardiovascular system differs from that during waking, since, in healthy subjects (Muzet), EEG and muscular reactions disappear from the third night of exposure, along with the subjective disturbance, whereas cardiovascular reactivity remains unchanged.

"These factors are even more apparent in children than in adults, and have several meanings: the first of these is that, without any suggestive counterpart (waking, disturbance), the noise signal is received by the nervous system since the latter triggers a cardiovascular response. The second meaning, which we attempt to interpret on the metabolic and anatomical level, is that as the noise only triggers these vegetative reactions, there must be structures that analyze and block part of its effects, preventing the signal from reaching non-specific zones, such as the reticular centers responsible for waking. Obviously, this represents a cost in terms of effort and energy, therefore leading to fatigue.

"This is where experiments carried out with animals are necessary. In the rat, electrodes were placed in the Locus Coeruleus, the node where essentially noradrenergic information is integrated, and in the Raphe Dorsalis, an essentially serotoninergic node, with numerous projections and varied regulations. Prepared in this way, the rat was left to sleep in an environment of regular sounds, resulting in an apparent difference between the behaviors of these two nodes. For a certain time, the response of the catecholinergic nodes was accompanied by waking and a reaction of orientation. This then disappeared and the animal went back to sleep, despite the noises. The response of the neurons in the serotoninergic node was very different. This was apparent, accompanied with that of the neurons of the Locus Coeruleus nodes, at the start of sleeping, and for each noise followed by a waking reaction. When the animal fell asleep in spite of the noises, and in contrast with what was observed in the catecholergic nodes, the response of these serotoninergic neurons continued without habituation for each noise during sleep with neurovegetative reactivity as an additional factor. However, the various functions of these nodes are involved in a large number of regulatory mechanisms, especially those of circadian rhythms and the cardiovascular system.

"What is more, according to postmortem studies carried out on depressive subjects after suicide, a decrease in serotoninergic neuron activity was often found in this structure.

"It should be borne in mind that in this general framework of passing from physiology to pathology, the high sensitivity to noise of depressive patients to nocturnal and diurnal noises can facilitate the appearance of their symptoms, and the fatigue felt by persons sleeping in a noisy environment, whether they are conscious of this noise or not".

2.5 Applications of the dose-effect relationship

The notion of dose-effect was defined at the beginning of this chapter. The following text by Jansen[24] gives his views on the different types of noise evaluation, a vision enriched by his 40 years of experience as a doctor.

"The physical (medical) well-being is the base for health. It is characterized by equilibrium of the physiological functions and functional systems of the human organism within standard deviations that have differentiated measures in resting and working situations. Diseases occur when measured values (blood pressure, combination of hormones etc.) exceed the area of standard deviations. Disease means an irregularity of physiological functions and the treatment of these irregularities. Healthy or ill reactions can be assessed by different methods. The upper limit of healthy reactions and the beginning of pathological reactions overlap so that a doctor always has to decide according to the harmfulness or harmlessness of a reaction in respect of disease. This type of assessment is a characteristic in curative medicine, whereas in social medicine the method of percentage assessment has to be applied.

"When establishing noise effect guidelines, two extreme positions have to be taken into account: (1) jeopardy of physical health (disease) and (2) complete health (high quality level of life and comfort). Between these two positions an area of decreasing health and increasing disease exists.

"These classifications of noise effects might serve as a base for a human oriented noise management.

"It is obvious that noise loads of LAeq = 30 dB(A) and Lmax = 45 dB(A) are representative for complete well-being which can be achieved only at a few places. LAeq = 75 dB(A) during daytime should be regarded as a tolerable limit for physical health. Long-term disturbances by noise inducing considerable annoyance were observed by levels of LAeq = 65 dB(A) to prevent negative psycho-physiological health effects.

"Below the guideline of LAeq = 65 dB(A) a great variety of psychological and social reactions (like speech interference, annoyance etc.) has been observed. But there is no evidence for causal relationships for negative health effects in the pathogenic sense. Nevertheless, requirements for noise reductions are necessary especially with noise sensitive situations or groups (e.g. learning in school etc.). The threshold for annoyance and communications was found at Leq = 55 dB(A) (0-20% annoyed) and at short time exposures of 55 dB(A) for 99% speech intelligibility. Noise levels during evening and night should be 5-10 dB(A) lower than during the day".

These guidelines proposed by Jansen are acceptable, except for the level of awakening reactions which seems too high. Personally we observed that awakenings occurred at 40-42 dB(A) Lmax indoors and we concluded that 50 dB(A) is a level which provokes awakening reactions. The WHO task force for noise suggested the level of 45 dB(A) should not be exceeded during the sleep.

Table 1: Graduation of noise effects on human health (Jansen[24]).

H (Healthy state):	complete well-being comfort high quality of life	LAeq = 30 dB(A) (in) Lmax = 45 dB(A) (in)
IHD (Increasing Health Disturbances):		
1	slight disturbances, negative influence on creative work, etc.	LAeq = 50 dB(A) (out)
2	threshold of annoyance vegetative reactions during sleep	LAeq = 55 dB(A) (out) Lmax = 55 dB(A) (in)
3	communication disturbances short time serious annoyance awakening reactions	LAeq = 55 dB(A) (in) LAeq = 65 dB(A) (out) Lmax = 60 dB(A) (in)
D (Diseases):	jeopardy ↓ hearing loss extra-aural overstrain	LAeq ≥75 dB(A) (out) LAeq = 85 dB(A) (in) Lmax = 99 dB(A) (in)

2.6 Combined effects

The beginning of the comprehensive analysis presented in section 2.3 emphasized that certain physiological responses to noise are not specific to noise stimulation. These responses of the organism can be facilitated or produced by other stress factors. Thus, regarding auditive problems, light has been shed on the role of high concentrations of carbon monoxide in the increase of ciliary cells damaged by noise (Young *et al.*[25], Fechter *et al.*[26]). An epidemiological survey on smokers exposed to noise during work showed more auditive difficulties than for non-smokers.

Several organic solvents are noxious for audition: toluene (Sullivan *et al.*[27]), styrene (Muijser *et al.*[28]), trichloroethylene (Rebert[29], Crofton[30]). However, Fechter *et al.*[26] was unable to show the ototoxic role of high doses of styrene in laboratory animals. The most recent studies carried out by Pujol at Montpellier and the team of Jorgen Fex at Bethesda, have contributed to knowledge on the role of neurotransmitters in the cochlea and we are starting to discover the intimate mechanisms of cochlea neurochemistry.

Many substances may play a role in neurotransmission. It seems that transmission of an auditive message between internal ciliary cells and the auditory nerve filter uses a glutamate or analogous substance. When precise knowledge of the role of the different neurotransmitters and their receivers is finally obtained, it will be possible to increase or inhibit them by synthetic substances. This has been underlined by Garcia[32].

Regarding the environment, people are frequently exposed to several sources of noise successively throughout the day. In urban environments, a person can receive noises from several sources at the same time. Rabinowitz[33] drew attention to the combined effects of noise and substances toxic for the auditory organ. The author emphasizes that "Certain antibiotics, in particular aminosides, cisplatine (used in cancer chemotherapy), loop diuretics and salicylates are therapeutic agents with significant ototoxic potential. The aminoglycosides (kanamycine, neomycine, streptomycine, gentamicine, etc.) and cisplatine can cause permanent loss of auditive perception (permanent increase of auditive threshold) with destruction of cochlear ciliary cells, while the loop diuretics cause temporary loss of auditory perception (temporary increase of auditive threshold). As for salicyslates, especially strong doses of acetylsalicyclic acid, they can trigger reversible tinnitus (humming, ringing, whistling in the ears)".

The author then describes the types of noise that can be measured in a hospital or clinic and the reduced defenses of hospitalized patients. "Concerning vegetative reactions due to noise, these have a threshold of from 11 to 30 decibels (dB) lower than that of persons in good health. Likewise, their aptitude to face noise generated stress is reduced. However, it is important to mention that nuisance to and disturbance of hospital patients are due more to the noises internal to the hospital than to environmental noises (road traffic, airplanes, etc.). Among the internal noises mentioned most are the noises of trolleys in corridors, nurses and neighboring patients, alarm bells belonging to the different devices attached to the patient and different communication devices (beeps, telephones, loud conversations, etc.). This can also have consequences for the hospital personnel and a study carried out in the United States has reported a relationship between noise levels and more frequent headaches (and demotivation) in nurses working in intensive care units".

It concludes: "After carrying out studies on medicines and ototoxic substances, it is apparent that more vigilance is required in the case of possible exposure to very intense noises, since effects harmful to audition can be heightened". However, in hospitals, sound levels do not reach the point where they lead to auditory problems, but it is certain that hospital patients constitute a group sensitive to noise, and that it would be preferable for hospitals to provide a certain level of acoustic comfort.

The complexity of the links between acoustic nervous stimulation and different human biological and psychological reactions, some of which we have just observed, lead to two conclusions:

- noise is not an electric current leading to a reaction proportional to the intensity of the stimulation, in this case acoustic. Triggering sensitivity varies and different noise levels can cause the same reaction, for example, of cardiac rhythm or awakening during sleep. Likewise, the same magnitude of a simple physiological reaction does not completely depend on the intensity of the stimulation that triggers it. The biological response to a noise comprises a certain amount of autonomy, and the same peak noise of a passing train, for example, can modify sleep by causing a slight reaction (transient activation phase), a change in the sleep phase or full awakening.
- consequently, this complexity leads to difficulty in proposing noise thresholds that can be considered as having a direct cause of an effect on health. Moreover, most works have focused on highlighting the effects of noise in a statistical manner, with conclusions in the form of probability expressed in terms of risk – thus we speak of epidemiology – rather than seeking to deepen understanding of the biological phenomena caused by noise.

Taking these observations into account, we will analyze the effects of noise on sleep, then detail the physiological and pathological effects reported by the main studies.

3 Sleep disturbance by noise

3.1 Introduction

People living in the vicinity of major roads, airports and other sources of noise maintain that their sleep is disturbed, that communications are impaired, that they are obliged to close their windows when listening to the radio or television and that the noise can give rise to psychological disturbances.

Various psycho-sociological inquiries concerned with noise have been conducted with a view to evaluating the different kinds of sleep disturbance and the frequency of its occurrence. Studies have shown how noise can affect the consumption of medicaments, particularly sleeping pills. Some inquiries have been concerned solely with sleeping difficulties.

Another way of evaluating sleep disturbance is to study the physiological changes induced by noise. For example, observations of sleeping subjects have been carried out both under laboratory and field conditions using electroencephalogram and electrocardiogram (EEG and ECG) recordings. The investigators were interested first of all in isolated changes in the EEG and ECG recordings associated with individual noises and then more recently with changes in the sleep structure itself in relation to the noise energy. The advantage of conducting physiological tests in situ is that the conditions that apply are more realistic in comparison with what can be reproduced in the laboratory, despite the fact that the attachment of electrodes to the scalp while asleep is somewhat unnatural. It is also possible to observe the effects of exposure to noise over very long periods of time, e.g. several years, whereas tests carried out in the laboratory are necessarily of limited duration. A study of the long-term effects of noise is of fundamental importance since, as in the case of sociological inquiries or those concerned with epidemics, this allows for the determination of the way in which the sleeping subjects adapt to the exposure by changing their behavior (e.g. closing windows, changes in sleeping periods), by modifying their dwellings (e.g. sound proofing of the façades of buildings and changes in the arrangement and uses of rooms), or as a result of physiological adjustments, which result in decreasing responses to the noise stimulus.

In what follows, a review is given of research which describes the nature and purpose of sleep, the techniques used for the evaluation and description of sleep, and the effects of transport noise in both its structure and quality. The aspects of adaptation to sleep disturbance, the use of medicaments to aid sleep and the use of physical indices to describe night-time noise are included.

3.2 Nature and structure of sleep

3.2.1 Sleep characteristics
There are three main states of existence which can be described as wakefulness (W), classic deep sleep (made up of four different stages labeled I to IV) and paradoxical sleep (PS). These different states or stages can be identified by observing several physiological changes occurring during sleep. These have been identified by Rechtschaffen and Kales[34] and also adopted as an international standard.

Other important parameters can also be considered such as heart and breathing rates, blood pressure and sweat gland excretion, etc., but this is not essential to an identification of the different sleep stages. Sleep is made up of a succession of cycles involving the different stages.

3.2.2 Physiological factors affecting sleep structure
Apart from quantitative and qualitative variations associated with the age of the subjects it should be noted that there are types of sleep that will have been

inherited or established at a very early age. A number of studies have resulted in subjects being classified as short sleepers (sleep durations of less than 6 hours) or long sleepers (sleep durations of more than 9 hours). The subjects concerned need to have slept for the applicable period of time for them to have the impression that they have had a sufficient or good nights' sleep. These differences are also associated with variations in the relative proportions of the different stages with respect to the total amount of sleep. The variations arising here must be regarded as sources of possible errors in cases of studies involving non-uniform samples of subjects.

In the case of an adult the average sleep duration amounts to approximately 8 hours and can vary from 6 to 10 hours or sometimes more. For a newly born baby, wakefulness and sleep alternate during the 24-hour period. The total duration of sleep amounts to 15 to 18 hours. Between 2 and 5 years, depending on the individual child, the monophase structure is acquired (i.e. awake during the day, asleep at night). The duration of sleep progressively decreases: 14-15 hours at 1 month, 11-12 hours at 1 year and 9-10 hours at 10 years.

In the case of old people (i.e. above the age of about 70) the duration of sleep is of the order of 6 hours, and it will be noted how there are periods of somnolence during the day and of wakefulness during the night.

3.2.3 Sleeping difficulties and normal variations in sleep
A brief reference to sleeping difficulties due to different circumstances needs to be made here in order to be able to make a distinction in considering those due to noise. These difficulties are concerned in particular with the sleep structure. Insomnia is very common and regarded by the medical profession not as a sickness in itself but as a symptom. The different types of insomnia can be distinguished according to when it occurs: difficulties in getting to sleep, waking up during the night or waking up too early. The general effect is a reduction in the amount of sleep.

Insomnia is a symptom of most mental disorders. Mouret and Vallet[9] have pointed out how the characteristic features of sleeping difficulties associated with a uni-directional depression is a reduction in the time preceding the appearance of paradoxical sleep and in the duration of slow sleep and/or frequent awakening.

3.3 Techniques employed in evaluating the quality of sleep

3.3.1 Questionnaire inquiries
All inquiries into the general effects of noise involve questions concerning sleep. Borsky[35], in a survey in Oklahoma, found that 42% of the people questioned referred to disturbed sleep due to noise. This was supported in a similar survey carried out by Lambert *et al.*[36].

Questionnaire surveys concerned specifically with sleep disturbances due to traffic noise have been carried out by Langdon[37] and Page[38], the latter being concerned with sleeping difficulties before and after the opening of a motorway. François[39] attempted to determine the effect of noise from airports on sleep while Fidell and Jones[40] evaluated the effects on sleep of canceling night flights at Los Angeles airport.

Some psycho-sociological inquiries have been concerned both with questionnaires and with observations of sleep behavior. For example, Rylander *et al.*[10] studied body movements in relation to sonic booms of a sub-sample of subjects from a larger population that had been involved in a previous inquiry.

3.3.2 Recordings made during sleep
It is much simpler to carry out recording procedures in the laboratory than in people's homes. However, the recording of body movements can be a useful way of studying sleep quality without the need for intrusive equipment. For example, Rylander *et al.*[10] used this technique to carry out tests on sleeping subjects in their homes when studying (like Ollerhead[41], Fidell *et al.*[42], Griefahn[43]) the effects on sleep of sonic booms. Due to some unexpected results from studies applying this technique, there is some doubt on its reliability.

The multichannel recording techniques that are employed are those that have been developed in hospitals for use in studying the sleep of sick people. The way in which the recordings are interpreted in order to identify the different stages of sleep was standardized in 1968. The different reactions that can be detected on taking account of this standardization include:

1. Changes in the EEG recordings
   a) Appearance or momentary disappearance of a standard electrical activity or momentary disappearance of delta waves.
   b) Appearance of a transient activity phase associated or not associated with a change in the stage of sleep (any change being in the direction of wakefulness).
   c) Occurrence of a more or less prolonged period of wakefulness.
2. Cardiovascular changes usually in the form of changes in the heart rate or peripheral vasoconstrictions.
3. Electrodermal reactions.
4. Muscular reactions ranging from a slight body movement to a change in body position.
5. Changes in breathing rate or amplitude.

The recordings can be made either in the laboratory (Lukas[44], Muzet and Erhardt[45]) or in the home (Globus *et al.*[46], Vallet *et al.*[47]). Working in this way it is possible to compare the sleep stages in a number of different noise situations and with a standard reference sleep structure, i.e. duration and rhythm of the different sleep stages in the subjects.

The investigators also considered isolated changes in sleep due in most cases to noise from particular sources such as trains, aircraft and lorries. This is a matter of transitory activities, changes in the depth of sleep stages and of periods of wakefulness. In these cases it is necessary to relate the number of sleep stage changes to their duration.

3.3.3 Morning questionnaires on sleep

Questionnaires have been employed in the case of laboratory tests where noise levels are varied during sleep and where the subjects are asked to give their impressions of their night's sleep on awakening. Depending on the nature of the experiment, subjects have been asked either to give a single indication of their sleep or to complete a questionnaire. The most well-known scale employed here is the Standford Sleepiness Scale (SSS). The questionnaires, in addition to being concerned with the general impressions of the subjects, are aimed at obtaining information that can be related to their physiological activities. The recollections of the subjects with regard to the time taken to get to sleep, the number of times that they awoke, and the number of dreams are compared with the indications given by multi-channel recordings that will have been made.

3.3.4 Morning performance and vigilance tests

In some cases the use of morning questionnaires is associated with fairly short duration psychomotor performance tests and sometimes with longer duration tests aimed at determining the vigilance during the course of the day on the assumption that disturbance to sleep due to noise will affect the performance and alertness of the subjects. The subjects used by Wilkinson[48] were made to complete the following four performance tests on the morning after their night's sleep:

1. Simple reaction time: the subject responds as quickly as possible to the random appearance of a three-figure number which starting from zero immediately increases until the subject stops the process on pressing a button.
2. Time of reaction to four choices: the subject presses one of four buttons which light up in a random manner.
3. Short-term auditive memory: the subject is presented orally with a list of 8 numbers which he has to write down (duration of test: 10 minutes).
4. Vigilance test: the subject listens for 1 hour to sounds of a half second duration and must detect a sound of slightly less duration.

The results of these tests give a quantitative indication of the effects of sleeping difficulties experienced during the preceding night.

3.4 Effects of noise on the structure of sleep

Investigations into the effects of noise changes on sleep demonstrate that noise from traffic and aircraft can be responsible for significant changes in sleep structures.

3.4.1 Psycho-physiological reaction to an increase in noise levels

Generally, an increase in the noise level during the night is associated with changes in the structure of sleep. Griefahn and Gros[49] noted that after a moderate increase in the noise level during the night there was a slight change in sleep, while Ehrenstein and Muller-Limmroth[50] observed that there was first a reduction in the amount of delta sleep followed by a decrease in the amount of paradoxical sleep. The results of a study carried out in people's homes by Vallet *et al.*[51] has shown that in real conditions the intrusion of noise leads to a reduction in the time preceding the appearance of the first period of paradoxical sleep and in the duration of delta sleep and an increase in the number of awakenings during the night. It also appears that irregular noise leads to more pronounced changes of stage than noise that appears after a uni-directional increase in noise level (Ohrström and Bjorkman[52]). There is a reduction both in the subjective appreciation of sleep and in psychomotor performance on awakening following exposure to noise at night. Results obtained by Metz and Muzet[53] show how, under laboratory conditions, subjects soon became accustomed to repetitive noise. It was found that their subjective reactions and certain recorded physiological changes showed a degree of adaptation while at the same time the subjects' cardiovascular reactions continued without showing signs of adaptation to the increases in noise. Similarly, the results of some of the investigations are particularly interesting in that they show how there was generally a lack of agreement between the way in which subjects became accustomed to variations in noise level. In these cases, the subjects' impressions of sleep quality indicated they were becoming accustomed to the noise whereas other physiological measures of sleep structure and, in particular, cardiovascular reactions did not indicate adaptation.

3.4.2 Psycho-physiological reactions following a reduction in noise level
Whatever tests are organized, statistically significant changes arise in both the duration of the sleep stages and the responses to isolated noises following moderate reductions (of 6-14 dB(A) Leq) in the level of noise during sleep. The results of three investigations also show that there is a positive correlation between noise level (Leq averaged over 1 minute) and heart rate whatever stage of sleep is involved. A reduction in noise level is usually followed by an increase in the duration of paradoxical and/or of delta sleep, the latter being little affected by isolated noises, and a reduction in the number of awakenings. Friedmann and Globus[54] investigating the effects of a reduction in noise level found that following a long period of exposure, the maximum rebound or increase in delta sleep occurred a week after the noise level was first reduced and steadied off after 1 month. Interestingly, a quieter night-time environment results in an increase in the amount of paradoxical sleep of old people and an improvement in the delta sleep of the youngest subjects. In addition to the EEG improvements their sleep has an improved subjective quality and the subjects have a better morning performance. This indicates that there is no physiological adaptation to night noise in the long term.

3.4.3 Accumulated sleep debt
A light sleep deficit each night provokes an accumulated sleep debt, and when the conditions of sleep change a clear increase in the deep sleep stages occurs. This acts as a counter-balance to sleep debt and is often called a rebound effect. A number of investigations have been carried out where the sleep structure of

subjects was recorded first of all in a normal noisy environment, N1, in a quiet one, and again in a noisy environment, N2. Wilkinson[48] noted that there were differences in the sleep structure recorded during the two noisy periods, the subjects being less disturbed in the second noisy period, N2, than they were in the first noisy period, N1. This was true for the total duration of sleep, for the light sleep stages I and II and for the deep sleep stages III and IV (delta). These results suggest that the quiet period results in a certain recuperation with a complete or partial elimination of the accumulated loss of sleep, the sleep in the N2 period being generally of a better quality than that in the N1 period. This supposition is, however, based on a limited amount of data and further experimental verification is required.

3.4.4 Factors affecting the assessment of sleep structure in response to noise

It should be noted at this point that measurements of sleep structure or physiological change are subject to considerable variation for a number of reasons. Generally, there are two main types of variation that need to be considered. These are related to individual subject variability and variations in the recording conditions between different investigations. Two major parameters, age and sex of the subjects, affect the reactions to noise. Young subjects (21-27 years) who are moved to a quieter room get to sleep on average in a shorter time, and experience a longer deep sleep, whereas older subjects (63-73 years), for whom deep sleep is physiologically shorter, experience a longer paradoxical sleep. The reduction in body movements is, however, the same for both age groups. Women are more sensitive to noise than men. This latter difference will be considered in more detail in the case of isolated noise (aircraft and lorries). It should be noted that no account has yet been taken of the two classes of subjects, namely short duration and long duration sleepers, which is another and very important cause of variation in sleep structure.

Finally, it is worth noting that Mouret *et al.*[55] indicated that the noise dose received by subjects during the day could have an effect on the quality of sleep at night. It is surprising to note that this important source of variation has not, as yet, been considered in studies of noise and sleep structure. Despite the various sources of interference with the sleep structure it should be recognized how noise can give rise to changes (distributed according to individual reactions) to the following sleep stages: deep (sleep stages III and IV), paradoxical sleep (length of time before the appearance of the first period and duration of each period) and periods of wakefulness (length of time before first getting to sleep and duration of wakeful period during the night).

The major problem remains that of interpreting the test results, when there are clearly distinguishable physiological changes resulting from noise exposure. It is not possible to disregard the possibility that noise is a hazard to health given our poor understanding of biological functions of sleep and our inability to attach any clear functional significance to change in the EEG signals. In conclusion, transport noise clearly disturbs the natural sleep process, and the consequent physiological disturbances are associated with a great deal of annoyance. Associated cardiovascular responses and otherwise unnecessary changes in certain physiological functions could lead to health problems, after very long term exposure to noise; however, there is, as yet, no clear indication that transport noise exposures are harmful to health in normal community settings.

3.5 Transitory disturbances to sleep due to isolated occurrences of noise

This section is concerned with isolated occurrences of noise due to aircraft, trains and road vehicles whose peak noise rises well above the background, e.g. noise of lorries in moderate or low flows of traffic.

3.5.1 Analysis of EEG recordings made in the laboratory

Two reviews of the literature concerned with work carried out up to 1975 and 1999 have been made by Lukas[56] and Ouis[57]. Among the criteria that can be employed in assessing disturbances to sleep described by Williams[58], two more important ones, namely awakening and changes of stage are considered here. The main external factor that needs to be taken into account with regard to immediate disturbance to sleep is the peak noise level. Tests carried out by Berry and Thiessen[59], Collins and Iampietro[60], Ludlow and Morgan[61], Lukas[62], Muzet *et al.*[63], and Thiessen[64] illustrated the effects of noise on 118 subjects aged from 1 month to 75 years who were exposed to over 8000 occurrences of noise, mainly from overflying aircraft, for a total of 769 nights. The coefficient of correlation between the percentage awakening with respect to the number of occurrences of noise and the noise levels, was r = 0.826, the peak level accounting, therefore, for about 68% of the variance in awakening reactions. According to Griefahn *et al.*[65] it can be seen that there were no awakenings for peak noise levels of less than 60 dB(A). However, awakening can occur below this level, particularly for old people or when the noise occurs during light sleep towards the end of the night sleeping period.

The relation between the absence of any changes of stage in the depth of sleep due to noise (zero reactions) and the level of indoor noise is another indication of disturbance to sleep whose use is strongly recommended by Lukas[62]. For the relation between the 'zero reaction' and the aircraft peak noise level L(A), the coefficient of correlation was r = 0.57. The data indicate that there are no changes of sleep indicated for peak noise levels of less than 37 dB(A) but the proportion of subjects for whom there was no such change falls to 44% for peak noise levels of 87 dB(A). Consequently, by limiting peak noise within bedrooms to 37 dB(A) most sleeping subjects will not exhibit any significant change in EEG reactions. However, controlling transport noise to such low levels is, at present, not a practical proposition. A laboratory investigation by Osada *et al.*[66] revealed that the time taken in getting to sleep increased with the level of train noise. For example, it was found that it takes 2-3 times longer for a person to get to sleep with peak noise levels of 60 dB(A) due to passing trains than it does in the case of a background noise level of 40 dB(A). Osada also found that the threshold noise level resulting in a person waking up amounts to 60 dB(A), and that the subject is kept awake for a longer period and the disturbance is regarded as more severe as the noise level increases above this threshold value. Finally, he found that the subject sometimes returns to a lighter level of sleep for peak noise levels above 40 dB(A) and very frequently does so for peak noise levels greater than 50 dB(A). Lukas[62] reported that sleep arousal thresholds are lower in women than in men. This was found to be independent of sleep stage or the type of noise stimulus (i.e. simulated sonic booms or flyover noise).

In more recent investigations it has been possible to prolong the period of observation in the laboratory and to make use of indications other than changes in EEG recordings. Muzet *et al*.[63] have recorded variations in heart rates and peripheral vasoconstrictions. In carrying out tests in the home it has been possible to relate the rates of reaction to different levels of noise over long periods of time. Thiessen[64] has demonstrated how sleep responses to noise (7 occurrences at a level of 65 dB(A) each night) changed over a period of 24 nights. The number of awakenings decreased significantly during the period of exposure while a change of stage generally appeared for each occurrence of noise. Vallet[8] has shown how this same distinction between awakenings and stage change frequencies seems to disappear after several years' exposure to noise. Jansen[6] has shown that noise levels of 55 dB(A) did not wake up his subjects although there were sleep changes and vasoconstrictions that were related to changes in the depth of sleep. Muzet *et al*.[63] studied the effects of the noise of individual vehicles on 26 subjects in the laboratory, with peak noises ranging from 40 to 65 dB(A) and 90 occurrences of noise per hour. It was shown that heart rate and vasoconstriction varied according to the peak noise level, particularly when the levels were between 60 and 65 dB(A). Examination of the responses for each 2 hour period showed that the subjects did not become accustomed to the noise during the course of the night. The tests also revealed that the cardiac and vasomotor responses were less pronounced in the case of old people. Regarding noise thresholds, it was found that responses began to appear for levels exceeding 50 dB(A) for children, 55 dB(A) for old people and 60 dB(A) for young adults.

Alternatively, Rice[67] has shown in a study of the effects of sonic booms on sleep disturbance that children were relatively unaffected when the peak external over-pressure was in the range 25-30 N/m² whereas 30% of the middle-aged population woke up as a result of disturbance, particularly in sleep stages I and II, and a very slight effect on δ sleep (deep sleep). However, the subjects' impressions of sleep quality indicated no significant improvement. Vernet[68] has studied isolated disturbances to sleep due to the noise of trains and compared *in situ* effects of train noise and that due to road traffic for levels of 70 Leq dB(A). It was found that the total number of individual disturbances was three times greater for lorries than for trains. The data shows that for trains there were no awakenings for peak noise level of less than approximately 50 dB(A) and even for peak noise levels of more than 70 dB(A) there was only a 2.5% probability of an awakening or a stage change. Peak noise level is not the only noise parameter which can be related to sleep disturbance caused by even an individual noise. Griefahn[69] has studied the effects on disturbance of the time interval between successive occurrences of noise. It was found that there was a maximum probability of an EEG effect when the interval between noise events was 40 minutes. This result is of practical importance in connection with the possible scheduling of night-time air or rail traffic. For example, it would be preferable to group the aircraft occurrences together provided safety considerations allowed rather than spread them out. Muzet *et al.*[63] demonstrated that an average passing frequency of 1.8 vehicles per minute (cars and lorries) gives rise proportionally to a greater EEG effect than a higher frequency of 4.3 vehicles per minute.

Ohrström has also studied the effects on sleep of noise level fluctuations. It was found that subjects in the laboratory where there was an overall noise level of 51.4 dB(A), were more sensitive to intermittent noise than to steady state noise. It was found, for example, that the percentage of body movements amounted to 16% for steady state noise compared with 22% for irregular noise. The type of environment where the noise is generated and the personality characterization of the population can apparently affect sleep disturbance in some instances. For example, Rylander *et al.*[10] established how a military population was much more tolerant of sonic booms occurring during the night than was the civilian population. Research conducted by Thiessen[64], Rice[67] and Muzet *et al.*[63] have shown how the age of the subjects can affect the extent to which their sleep is disturbed by noise. All these different factors are taken into account in what Fidell *et al.*[70] refer to as the detectability of identifying a noise (where we are referring, of course, to a clearly identifiable noise and not noise due to continuous, dense traffic). The vulnerability of sleep is very dependent on these different factors. Pearsons[71] pointed out the difference of results obtained in laboratory and in field studies (cf. section 3.5.2 below).

3.5.2 Recent investigations

Surveys carried out during the 1990s highlighted the significance of the impact of night-time noise, and high demand from populations exposed to noise to conserve the quality of their sleep.

Since the last congress in 1993, in Nice, France, research into sleep disturbance by noise basically falls into three categories:

– experimental research aiming to describe the physiological effects of noise on sleep. In addition to conventional parameters used to describe sleep (electroencephalograms, electrocardiograms, occulograms, breathing and actigraphics), biochemical scales are now used to explore noise-related modifications. All psycho-sociological field and laboratory surveys are also included in this category as are epidemiological studies demonstrating sleep disturbance after effects.

– research which aimed to determine the noise levels sleepers can tolerate by re-working data collected in previous studies or by obtaining new data specifically for this purpose.

– reviews of existing research, usually to prepare regulations, such as the often stringent recommendations of the World Health Organization and national standards.

The scientific community was surprised by the results of a vast survey carried out by the British Civil Aviation Authority[41]: "for outdoor event levels in the 90-100 dBA SEL (80-95 dBA Lmax) range the chance of an average person being wakened is about 1 in 75" even if the most sensitive subjects were disturbed twice as much as less sensitive subjects. Despite attempts to understand what variation factors could create such a difference from the findings of other studies, it was simply observed (Latham[72]) that average insulation provided by the windows of homes situated around British airports was, on average, 36 dB(A) which attenuated Lmax levels in bedrooms by 44 to 59 dB(A).

But no one knows if people who live around airports in Great Britain sleep with their windows open! A second aspect of the inclusion of noise in sleep disturbance studies is understanding the extent of multiple exposure. In 1978, in a brilliant experiment using twins sleeping in laboratories with an identical low background noise, Blois *et al.*[55] showed that the amount of noise to which people were subjected during the daytime determined the extent to which their sleep was disturbed. This daytime noise memory effect on night time sleep, confirmed by Frusthorfer[73], reveals the necessity for scientists - and law-makers - to include noise from all sources and even noises which people do not complain about. People living near roads are a good example.

This body of research describes the primary and secondary effects of noise on sleep since 1993.

Primary effects are assessed by arousals which are observed as a response to intermittent noise (EEG analysis):

– by the effects of noise on cardiac rhythm, including arrhythmia during sleep,
– the analysis of urinary catecholamine secretions which gives further insight into the activity of the sympathetic nervous system.

Secondary effects of noise on sleep are examined in the light of:

– modifications to the immune system, changes in the secretion of some hormones, particularly growth hormones. The Japanese school opened this path by studying the influence of noise on the size and weight of new-born infants,
– the level of blood cholesterol and risks of chronic cardiovascular disorders.

Despite the broad panorama of investigative methods, the overall number of experiments fell between 1993 and 1998 and the research published concerned relatively small populations. Carter[74] studied the effects of road traffic noise on the cardiac rhythm of seven old men and showed the effect of noise on four subjects afflicted by cardiac arrhythmia during some sleep phases. Two of the four subjects with arrhythmia presented a significant response during phase four sleep to a single loud indoor noise event.

Carter *et al.*[75] attempted to demonstrate relationships between road traffic noise, modifications in slow sleep and immune response in two groups of shift workers. Some of the results are significant and others are not.

This type of research requires large populations to minimize the impact of multicolinearity and relatively large research teams, which implies significant long-term funding. To some degree it is an adaptation to a new situation in noise-sleep research which was initiated by Ollerhead *et al*.[41], using a simpler method than EEGs. Actimetrics are somewhat more basic than conventional polygraphics and enable investigation of larger sleeper populations. However, as Griefahn *et al*.[76], showed this method still needs to be validated on a general, statistical and individual level.

Carter[77] prepared a broad-spectrum review describing the different effects of noise on sleep classified by experimental conditions (laboratory or field), noise type (continuous or intermittent) and emergence to background ratio. The effects of noise on sleep are included in a comprehensive manner, including arousals, phase changes, awakenings, phase durations and appearance phase latencies together with physiological measurements (cardiac rhythm, actimetrics) and after effects. Carter's objective was to determine physiological modifications which have a strong probability of inducing effects on health such as the cardiovascular system and immune response system at some time in the future.

The review prepared by Pearsons[71] uses data from 21 studies to propose a dosage-response function for the effects of noise on sleep. The author observed a significant difference between the results obtained in laboratories - where the noise effect is high - and the results obtained in the field in which subjects seem to become accustomed to the impact of noise.

The influence of noise on the sleep function depends on noise and sleep modification types as well as the noise source, the background noise level, the length of the study and the sex of the subject. It was not possible to validate a model for sleep disturbance by noise using data sampled from the main studies. It should be noted that this review basically addresses the effects of separate noise events (aircraft and trains) and does not include the effects of continuous noise on sleep structure.

To assess the nocturnal aircraft noise adaptation a total of 16 airport residents were studied over a period of 40 nights by Maschke *et al*.[78]. The test persons slept in their own apartments and were exposed to nocturnal aircraft noise. During this period, 32 takeoffs and landings with sound levels of Lmax = 65 dB(A) were simulated electro-acoustically. The night urine discharge was collected and analyzed. The time curve of cortisol excretion shows that no uniform reaction to persistent nocturnal aircraft noise exists. On the contrary three reaction patterns can be distinguished. The cortisol excretions exhibit rhythmic fluctuations as expected. At the end of the study the cortisol excretion exceeds the normal medical range. The initial reaction to the added nocturnal aircraft noise is marked; however the cortisol values remain within the normal medical range. The second figure (four test persons) shows a pronounced initial reaction followed by a decreasing cortisol excretion trend. The initial reaction clearly exceeds the normal medical range. The cortisol values of the first week correspond approximately to the results of the Berlin field study. The third figure (five test persons) shows the cortisol excretion barely changes. The initial reaction is slight. The weekly rhythm of cortisol excretion predominates.

Several experiments conducted by Kawada and Suzuki[79] revealed certain characteristic changes in the rapid-eye-movement sleep stage in response to noise exposure. Continuous and all-night exposure to noise first decreased the percentage of the rapid-eye-movement stage at Leq 45dBA. This decrease occurred earlier than the changes in any other sleep parameters studied. In contrast, the threshold of shifts from the rapid-eye-movement stage in response to intermittent noise was higher than thresholds noted for slow-wave sleep or stage-2 sleep. The authors concluded that these results indicated that a silent environment is required to maintain the duration of rapid-eye-movement sleep, even though rapid-eye-movement sleep is stable and is not influenced easily by noise.

3.6 Adaptation of sleep to noise: physiological habituation

A number of changes occur with continued exposure to noise. First of all there is physiological habituation in terms of a decrease or even the disappearance of the reactions of the human organism to the noise occurring during sleep. In the case of aircraft noise, it has been shown that the duration of exposure to noise plays an important role in the degree of sleep disturbance. It has been found, for example, that the number of awakenings and changes of stage decrease with the number of nights of exposure to noise. It is considered that the sensitivity to noise is affected at a very early stage, e.g. by the noise experienced by the unborn child. Ando and Hattori[80] have demonstrated that children born to women who were exposed to the noise from an airport during the beginning of their pregnancy subsequently woke up less frequently than children born to women who were exposed to the same noise only during the second half of their pregnancy. Muzet[7] has noted how, during the course of time, differences arise between the subjective appreciation of sleep which soon increases following the beginning of exposure to noise, the EEG responses which decrease only slightly and the cardiovascular responses which remain fairly pronounced. These differences become even greater in the case of very long periods of exposure to noise. Fidell and Jones[40], as a result of conducting an inquiry using questionnaires, found that people living in the vicinity of an airport were not aware of any improvement in their sleep despite the fact that night flying had ceased, while Friedmann and Globus[54], on recording the physiological responses of subjects in the same area at the same period of time, found that there had been a definite improvement in the quality of sleep. For example, it was found that the amount of deep sleep had increased from 12 to 17% one week after the cessation of night flying. The same difference was reported by Vallet and François[81], who concluded that physiological changes in sleep quickly follow changes in noise level while changes in the degree of annoyance follow much more slowly. This both demonstrates and confirms that the degree of annoyance expressed with regard to noise depends largely on factors other than the noise itself. In addition to this it is known that about 10-20% of the population say that they sleep badly for reasons other than exposure to noise.

3.7 Noise and consumption of medicaments to aid sleep

If it is accepted that an individual in good health has no need of medicaments then the consumption of medicaments by the population exposed to noise can represent a quantifiable effect of this nuisance. For example, studies carried out in Ontario and the Netherlands showed that there was a significant increase in the issue of medical prescriptions to people living in noisy areas when compared with a control group living in a quieter area. The total purchases of medicaments for high blood pressure per year increased for people living in the vicinity of Amsterdam airport and in proportion to the activity of the airport whereas there was no such change in the case of a reference group of people who were not exposed to the noise. With regard to sleeping difficulties associated with the noise of motor vehicles in urban areas, it has been found difficult to dissociate noise effects from those due to age, anxiety, pain and insomnia. According to Langdon[37] noise was not considered as a major factor in the consumption of medicaments such as sleeping tablets. Alternatively, Lambert *et al.*[36] suggest that there is a low correlation between the consumption of sleeping pills and traffic noise. The use of earplugs is not considered in these studies, although Wehrli and Wanner[82] found that the proportion of people making use of some form of earplug varied from 0 to 14% compared with 0-4% for the consumption of sleeping pills. The higher proportions of people using earplugs was associated with the higher noise exposures in the range considered. The results of a survey carried out by Relster[83] in Denmark show that in areas where the Leq levels are high, 69-78 dB(A), some 19% of the population compared with 12% in a reference area where the levels are lower, claim to have consulted a psychiatrist or a psychologist during the past 5 years. It was further established that 4% of these cases were referred to a psychiatric hospital whereas only 2% were referred from the quieter area. It is also claimed that 25% of the population exposed to high noise levels regularly took tranquilizers whereas only 17% took them from the control group.

Vallet *et al.*[84] carried out a pilot epidemiological survey, which in effect it was carried out by medical doctors around Paris Roissy airport. This feasibility study consisted of a comparison between patients exposed to aircraft traffic noise and those not exposed; noise levels are expressed using computed contours. The exposed group is similar to the non-exposed one, according to the main socio-demographic parameters and types of housing. Simple comparisons and a logistic regression analysis have shown trends of more frequent illness among exposed people.

A higher number of patients experiencing digestive spasms, excessive perspiration, or expressing a strong feeling of fatigue were found in noisy areas. More objectively, a significant increase of the use of tranquilizers, in addition to medicines for neuro-psychiatric problems, or anti-acid and anti-ulcerous drugs, was observed in areas highly exposed to noise (Odds Ratio 0.48 to 0.55). No difference was noted for the consumption of pain-killers or for sleeping tablets. Curiously, sleeping problems are relatively more frequent in non-exposed areas. This could be linked to the effects of a widespread prescription of neuro-psychiatric drugs in these noisy areas. The occurrence of blood pressure problems is a little higher (34% instead of 31%) amongst patients from noisy areas, without being significantly different. On the other hand there is a significant increase in the number of doctors prescribing sick-leave in the noisy zones (1 sick-leave: 27% ≠ 17%, 2 or more sick-leave: 25% ≠ 19%).

The analysis by logistic regression analyzed in detail 10 types of health effects: some of them are variables built by a combination of 20 items (psychosomatic troubles). The major effects of the exposure to aircraft noise are listed in the table below.

Table 2: Health effects according to aircraft noise levels.

Health effects	Odds Ratio		Relationship
	High noise level	Moderate noise level	
Complaints against noise	10	12	Very high
Anxiety for future	4.2	2	Strong increase
Prescription of tablets for neuro-psychiatric problems	2	1	Increasing with high noise
Sleeping pills consumption	2.1	1.1	Patient influence
Sleeping problems	1.7	3.7	Higher in the moderate noise zone
High blood pressure	0.9	0.6	No link

3.8 Sleep disturbance and performance the following day

It has been shown that sleep adaptation can arise during the course of successive nights to exposure to noise. However, even if there is a certain reduction in the indications of physiological disturbance, the performance of subjects during the day following a night's exposure to noise is still affected. Statistically significant increases in motor reaction times were noted by Vallet[8] and more particularly by Wilkinson[48]. The latter found that the unprepared reaction time to a single stimulus in the case of a multiple choice test was shorter following relatively quiet nights. However, performances involving the short-term memory were not affected by changes in the noise environment. The test results as a whole confirm the common observation that people sleep badly in the presence of noise and it is, therefore, necessary to consider the possibilities of establishing noise indices which can be used to determine criteria for the protection of sleep.

3.9 Night noise indices and sleep disturbance criteria

Some countries make use of noise indices specifically to qualify evening or night noise. They take account of such periods by making use of a composite index where the noise energy received during the evening and the morning is combined after appropriate weighting with that received during the day. The use of such an index can be considered to be unsatisfactory, since in practice it is incapable of taking sufficient account of the characteristics of the night noises that interfere with sleep, especially the maximum levels of noisy events. However, the development of noise indices that do not cater for both sleep disturbances and general annoyance require much better understanding of the parameters of the transport system responsible for the noise characteristics affecting sleep. Lamure[85] has reviewed the problems associated with developing noise indices related to sleep disturbances. Reports issued by various international organizations have shown that there is considerable interest in the subject of protecting sleep and the process of recovering from physical and nervous fatigue as a result of sleeplessness:

1. The European Communities Commission considers that a night-time Leq level of 30-35 dB(A) or below within buildings and peaks of 45 dB(A) or less do not affect sleep.
2. The Organization for Economic Cooperation and Development provisionally recommends the adoption of the following Leq levels in member countries: 35 dB(A) in the case of light sleep and 50 dB(A) for deep sleep.
3. The World Health Organization recommends an internal Leq level of about 30 dB(A) during the night (1999).

However, this latter recommendation, based on the results of tests carried out in the laboratory, was considered to be "unduly strict by" Large[86] The difficulty of determining the amount of traffic and consequent noise occurring during the night again raises the question as to whether limiting noise occurring during the day will not automatically result in a corresponding reduction in that occurring during the night.

It is generally accepted that the difference between Leq levels for noise occurring during the day and night in the vicinity of major road traffic routes is a limit of 10 dB(A). This implies that the application of a limit of outdoor Leq = 60 dB(A) for the period from 06.00 to 22.00 hours will automatically result in a limit of Leq = 50 dB(A) for the night period. This would appear to be sufficient to protect the general structural quality of sleep, except in the case of isolated occurrences of noise.

Maurin *et al.*[87] however, has shown that the difference between day and night noise levels is of a rather variable nature. It was found that for a sample of about 75 subjects (24-hour periods), the difference between the Leq value from midnight to 05.00 hours and that from 08.00 to 20.00 hours varied from 8.2 to 13.5 dB(A).

Differences between day and night levels for average noise profiles in the case of different classes of road are also found to vary over wide ranges. Specific noise profiles for which the day/night difference in levels amounts to less than 6 dB(A) are generally associated with minor roads. Small differences in these levels are to be expected in extreme cases of roads with heavy traffic, and if the difference falls below 6.8 dB(A) then the protection provided against exposure to daytime noise will not be sufficient to ensure protection against exposure to that occurring at night. A similar problem arises in the case of physiological responses. Mouret and Frusthorfer have shown that noise experienced during the day can result in disturbances to sleep structure the following night. This finding would appear to support the adoption of noise indices that are not solely related to specific night-time periods, such as the day/night level DNL or L10 (18-hour) (the 18-hour period is from 06.00 to 24.00 hours).

Several indices have been developed which include special weighting factors applied to night-time noise levels. For example, the day/night index referred to above is given by the formula:

$$LDN = 10 \log 1.24 \ (15.10 \ LD/10 + 9.10 \ (LN + 10)/10)$$

where LD = Leq for the period 07.00-22.00 hours and LN = Leq for the period 22.00-07.00 hours.

Other frequently used weighted indices are the CNEL index (community noise equivalent level) which includes a 5 dB evening weighting for the period 19.00-22.00 hours, the German Ltan index which is similar to CNEL but extends the period covered from 07.00-22.00 to 06.00-24.00, the French index and the British NNI (noise and number index). In the latter case the acceptable values of NNI differ for the day (50) and the night (30). A directive is now being prepared by the European Commission. An indicator similar to CNEL has been proposed by a specific working group, and finally adopted in July 2000.

Ollerhead[88] and Rice[89] support the use of CNEL, since it takes account of the noise during the period of the evening in which the maximum amount of annoyance often occurs. Two surveys have supported the use of Leq averaged over the period 23.00 to 07.00. In particular, Brooker[90] shows that total sleep disturbance increases slightly at higher Leq levels. Walker and Diamond[91] discuss the influence of background noise levels at night. Proposals for a threshold or criteria value for sleep disturbance have been made by referring to individual sleep response to noise, or to a percentage of the population affected by noise-related sleep disturbance. Rice[89] suggests that a minimum of 25% of the population should be treated as a limiting value when assessing sleep disturbances associated with noise sources. Schultz[92] has compared the disturbed sleep reported by a variety of surveys of different sources of transport noise.

He shows that the average LDN level that appears to disturb about 25% of the population is approximately 68 dB(A). If we presume that the difference between day and night values of Leq is from 5 to 10 dB(A), then this would suggest that a night-time Leq of 61-62 dB(A) at building façades would be roughly equivalent to 68 dB(A) during the day. This level represents a much higher level than those recommended by the EEC, OECD and the WHO and clearly reflects the difference between criteria based on sleep structure reactivity or physiological change and criteria based on people's subjective opinion of sleep quality.

Rucker *et al.*[93] also reported the results of studies of disturbances to sleep based on subjective surveys. The figures do not give any clear indication of an acceptable limit to noise level. According to the data, however, the sleep of 25% of the population is disturbed when the external Leq level (22.00-06.00 hours) is 54 dB(A), which is substantially lower than that found by the authors and indicates a higher external level than that recommended by the EEC, OECD and WHO. It is more difficult to postulate noise level criteria for sleep disturbance when the noise stimulus contains isolated occurrences. However, in certain situations, the use of one Leq value for a particular period of time dispenses with the need to consider the relevance of peak noise levels. For example, for road traffic noise, the existence of a high vehicle flow at a moderate or long distance from the reception point will not generate significant peaks in the noise level signal and, consequently, the use of an index based on the scale of Leq is probably sufficient to characterize the risk of sleep disturbance. In other circumstances, e.g. low traffic flows, a reference to peak noise levels is needed, although peak levels alone do not adequately explain sleep reactions. For example, for traffic noise exposures under quiet conditions, the average peak noise level causing an EEG effect may vary from about 42 to 44 dB(A) while that causing the same effect in a noisy situation typically ranges from 50 to 53 dB(A).

It can be concluded that the peak level of an isolated noise is not sufficient to take into account time dependent sleep reactions and, consequently, it is necessary to include some measurement of the general ambient level in addition to the peak levels. This general noise level, to which background noise makes an important contribution, may come from sources other than transport and is hardly ever taken into account in studying the noise from aircraft and trains. In laboratory tests Lukas[44] considered that the critical noise level for awakening as indicated by EEG responses was 60 dB(A) while Muzet et al.[63] observed that there was a definite cardiovascular response for noise levels of 50 dB(A), but that changes of stage occurred for noise levels of 45 dB(A). Vallet et al.[94] found that people who had been exposed to aircraft noise for 5 years did not wake up for peak noise levels of less than 55 dB(A) and that there were no changes of stage for internal noise levels of 52 dB(A).

To inform the relevant authorities of the problems raised by the night-time operation (freight transport) of Maastrich Airport (NL), De Jong[95] synthesized information resulting from a study on the effects of isolated noises on acoustic comfort during sleep. Recommendations for an acceptable threshold were for sound levels of 40 dB(A) inside bedrooms with peak levels lower than 55 dB(A) when the number of flights does not exceed twenty per night. If this number rises to about thirty, the peak level of each flight should be lowered by 5 dB(A). These recommendations are completed by attention given to sensitive groups, representing 25% to 30% of the population, for whom additional allowance in respect of housing insulation must be made.

This issue of insulation is raised by Tullen et al.[96], who opined that reinforcing apartment wall insulation to improve sleep quality was ineffectual. This assertion, which runs counter to the trend and results provided by all the other studies, points towards noise acting in an all-or-nothing way, which is quite possible on a theoretical and even experimental level. Nonetheless, the assertions of this team, which used as the disturbance criterion the cardiac reactions within the four seconds preceding and succeeding the peak noise, require modulation.

Eberhardt[97], in a synthesis drawn up for Nordic countries, noted that 20 noises from trucks with a peak level of 55 dB(A) caused considerable changes to sleep and he suggested selecting a peak threshold of 50 dB(A).

In the same ministerial report, regarding subjects in good health, Ohrström recommended against exceeding a threshold of eight noises per night of over 60 dB(A) and 40 noises with a peak level of 45-50 dB(A). For the comfort of sleep, it is advisable that peaks do not exceed 45 dB(A).

Nordic authors, including Rylander, unanimously reject the Leq as the energy index used for recording noise-related nocturnal disturbance, without having examined the correspondence between the peak levels, on the one hand, and the resulting acoustic energy, on the other.

As for Rabinowitz[98], he considers that sleeping problems occur for a night-time Leq on building façades of 55 dB(A).

This problem has been dealt with by several studies (Pearsons *et al.*[99], Griefahn[100], Vallet[8], Ollerhead *et al.*[41]). Pearsons' studies focus on the use of the Leq and the sound energy level peak SEL, whereas the three others constitute an attempt to give a dose-response curve as a function of the number of events and the peak level. Thus Griefahn, after examining a large number of existing works, proposed a model according to which the sleeper should not be subjected to more than 10 noises per night with an average level of 54 dB(A) inside the bedroom.

Using the results collected from the homes of persons living near an airport, and the model mentioned above, Vallet suggested that 90% of awakenings could be avoided if there were no more than from 10 to 20 noises of over 48 dB(A) inside the bedroom per night. By contrast, Ollerhead considers that an external noise of 90 dB(A) SEL (80 dB(A) Lmax) has little effect on sleep and that for external loud noises (80 to 95 dB(A) max) the average possibility of awakening is 1 in 75. These observations differ substantially from all the other results in the literature. What is more, examination of length of residency does not shed any light on these results. We note, however (Latham[72]), that a closed window in Great Britain provides an average insulation higher than 36 dB(A), resulting in an internal level of 49 dB(A). This makes the different research results more compatible with each other and we therefore support the conclusions which advise that peak noise levels should not exceed 45-50 dB(A) more than 15 times per night. Kuwano *et al.*[101] has studied the effect on sleep during a simple and clever experiment. Continuous road traffic noise was found to disturb sleep at a level of 45 dB(A) Leq, but not at 35 dB(A).

4 Pathological non-auditory effects of noise

It is difficult to disassociate noise from all the environmental factors specific and external to the person. Noise does not occur in a vacuum, either in biology or in physics. To understand how this group of non-auditive phenomena occurs, it is necessary to consider the different aspects of the notion of stress.

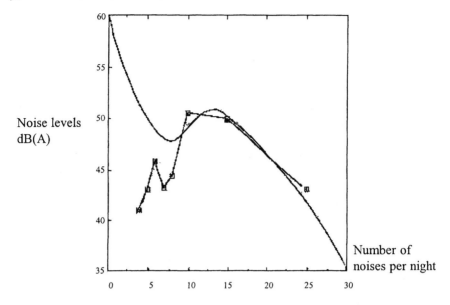

Figure 4: Disturbances according to the number and noise levels per night.

4.1 Stress

4.1.1 The notion of stress

Mouret and Vallet[9] describe stress as follows: "It is both a very popular word and, like all concepts, very difficult to define. It is used in a large number of areas, such as physics (synonymous in this case with tension: the effect of a force exerted against a resistance), biology and human sciences. Thus it is a word that, while remaining a metaphor, is considered scientific.

"Stress is above all a response made by the organism. This implies the existence of a stimulus that can be defined in an overall way as any activity in the environment that leads to a response, of any nature whatsoever, from a given individual, this stressful stimulus not necessarily being the same for everyone.

"However, living beings, and human beings in particular, are very sensitive to even very small changes in their environments. This leads to considering stress, on the basis of the notion of stimulus-response, as not only being the stimulus or only the response, but the entirety of this process, including what they involve, i.e. interaction.

"If we content ourselves with these definitions, it is obvious that they correspond very well with a physiological phenomenon, whereas the notion of stress is more often related to pathological phenomena.

"In fact, in the physiological process the interaction corresponds to a situation in which balance is restored rapidly after a transient modification of "homeostasis". In parallel with this is the notion of pathology, i.e. that of a lasting modification with regulations centered on other functional levels.

Obviously, defining limits between physiology and pathology is no simple matter, since they call into play notions that are either diachronic, or which still belong to these two all encompassing modes. The problem is even more complicated if one takes into account the definition of health as given by the WHO. Despite these difficulties, studies on stress can be classified into four main categories:

- Selye[23]: general syndrome of adaptation.
- Studies by Friedman and Rosenman[102], etc., that establish the statistical and epidemiological foundations in which the concept of stress is formulated.
- A group of approaches more difficult to summarize, but which can be classified in a psycho-sociological framework, covering both animal and human experiments in which, by using non-physical means, the environment plays a role leading to pathological effects.
- And lastly, all the more fundamental studies linking stress to the activation of the amine, peptide and neuro-endocrine systems, without mentioning behavioral responses and individual typologies".

4.1.2 Stress and general adaptation syndrome
Mouret and Vallet continue:

"This syndrome was discovered by H. Selye following experiments qualified as the pharmacology of dirt.

"After several intraperitoneal injections of placental and ovarian extracts, Selye observed the following in the animals treated:

- an increase in volume of the adrenal gland,
- atrophy of the thymus, spleen and the lymph glands,
- gastric ulcers.

In every case, whatever the products injected, all these changes occurred together and never alone in isolation.

"Therefore Selye considers that this is a non-specific reaction that evolves in three stages, which can correspond completely with what happens with noise:

- initial reaction of alarm manifested by acute manifestations,
- stage of resistance characterized by the disappearance of the signs of alarm,
- stage of exhaustion with complete loss or resistance.

"This group of symptoms permits considering the characteristics of the general adaptation syndrome as representing a non-specific effect. It is very important to note that this is a non-specific response by the organism to any stress, whether due to pleasant (eustress) or unpleasant (distress) phenomena.

"For Selye, this is a phenomenon of the organism reacting to stimulations, which, except in his most recent articles, are never psychological, in the form of disturbance and deregulation of homeostasis at the moment when the entire organism is occupied by managing the stress and emergency.

"The morphological triad of stress, based on the results from animals subjected to severe stress has many implications. Moreover, certain manifestations of stress can be recognized not only by structural alterations but also, and sometimes only, by functional aberrations such as modifications of behavior, responses to certain medicines and certain immunological stimuli.

"As for hormonal modifications used as evidence of stress, it should be remembered historically that they have been described for urinary catecholamines, glucocorticoids (Cortisol, Cortisone, 17OHCS), and certain androgens and estrogens. However, this evidence differs as a function of time and circumstance (the increase of urinary catecholamines is generally more transient than the increase in the rate of corticoids)".

These notions, already mentioned in section 2.3, are rarely associated with research into noise, apart from work done by Maschke *et al.*[78] (cf. section 3.5.2).

4.2 Epidemiological studies on the effects of noise

Repeated stimulation of the endocrine and nervous systems for several years increases the risk of chronic problems occurring in the persons exposed. Noise can also cause different individual responses: change of blood pressure, increase in cholesterol, and the appearance, for example, of mental anxiety and depression. Other stress agents cause the same reactions in persons exposed to noise and these other agents, to such an extent that it is difficult to isolate the effect of noise alone. Therefore it is necessary to carry out surveys on very large samples of the population, which permit controlling factors other than noise. As will be seen in the examples below, epidemiological surveys highlight, separately, the higher frequency of cardiovascular, metabolic and anxiety-depression problems among others. Each effect of noise on health shows a statistically higher occurrence in areas exposed to noise and sometimes a significantly higher risk. To our knowledge, there are no survey results that provide an overall balance of the effects of noise on health, and the aggregation of all the different elementary effects. From this, we conclude that audits of the impact of noise on health are under-estimated. Likewise, these surveys, which sometimes stretch over ten years, according to how they are designed, only take into account the survivors, i.e. the people who still live in the area, by choice or necessity. It has never been possible to interview persons who have moved away from noisy places, in order to evaluate the role of noise in the decision to move or the acceptance to pay more for a residence in a quieter place. This radical adaptation to a noisy situation by part of the population, *a priori* considered annoyed by noise, minimizes the impact of noise on the inhabitants, and on health related aspects.

Fluctuating noises are difficult to assess by the extremely precise methods applied to noises of short duration like those used in neurophysiology and psychoacoustics. When a noise fluctuates at a given place and when the same person is exposed to several noise sources that fluctuate over time, it is not possible to use the same methods to assess impacts. Epidemiological studies aim to demonstrate the effects of noise on health. Apart from the impact of noise on sleep, which is often of short duration, epidemiology deals with non-auditory effects on health and the most convincing results appear for high noise levels at the workplace. Noise can, for example, cause high systolic or diastolic (SBP, DBP) blood pressure.

Table 3: Epidemiological studies on noise.

Authors	Date	Exposure levels	Exposure duration	Results
Zhao	1991	75-104 dBA Leq (no ear protection)	Complete life	Very significant effects of noise levels
Lang	1993	85-100 Leq A	>25 years	Effects of length of exposure
Fogari	1994	75-100 Leq A		≠ significant :<80 Leq A and >85 Leq A
Talbott	1993	≤82 - ≥89 Leq A	> 15 years	Effects for 40-49 age group
Hirai	1991		10 years later	No effects
Hessel	1994	<80 ->100dBA	27 Years	No effects
Kristal	1995	<65 ->90 dBA		No effects on blood pressure
Boneh				Effects on heart rate in women

Obviously, it is essential to find out why four studies show that exposure to loud noise for long time periods has significant effects, whereas three prove the contrary. These studies are so massive that it is very difficult for research teams to consider noise exposure levels because often they have to be reconstituted for several thousand people. This is the first but not the only comment on methods. Thompson[110] wrote a genuine lesson of epidemiology about the non-auditory health effects of noise, explaining the concept of risk confounding factors, types of studies, etc. In her updated review, the role of annoyance due to real working noise, or to environmental noise levels, is considered as a relevant factor. Several investigators have suggested that non-auditory noise effects may be more closely correlated with suitable subjective noise parameters than with the real sound level. In the Berlin population-based case-control traffic noise study, subjective work noise assessments were compared to the median of measured sound levels in a sample of subjects, making it possible to express risk by noise levels. The risk for myocardial infarction (MI) at noise levels of 79-99 dB and >100 dB were 2.0 and 3.8 respectively, after controlling for multiple confounders. In a community survey Lercher and Kofler[111] found expressed annoyance with work noise had a slightly significant effect on DBP, but a non-significant effect on SBP. Apparently, no attempt was made to relate noise annoyance to noise level. Recent findings from studies of physiological effects of relatively low environmental noise levels point to the importance of ambient noise as a widespread stress factor. The recent Luebeck study indicates a slightly higher prevalence of hypertension in men living in high traffic areas compared to low traffic noise areas (odds ratio 1.3). Whereas blood pressure effects were not observed in another well controlled cross-sectional study, significant associations were found between noise and heart disease risk factors, including platelet count, glucose, plasma viscosity and total triglycerides.

After reminding us of the poor quality of the first epidemiological studies, the author acknowledges that studies have been better prepared since 1980. Recent results obtained in surveys in Caerphilly and Speedwell and in a case-control survey are subtle or show a trend. The risk of cardiovascular pathology increases when workers are exposed to high road noise levels (66-70 Leq A compared with 51-55 and categories 71-75/76-80 compared with 51-60) (Babisch[112]). A review is given by Thompson[16] on ischemic heart disease, blood pressure changes and effects of noise on the fetus. Case-control epidemiological works imply having a reference situation where noise levels are low. For people exposed to aircraft noise this type of survey is difficult to implement, as aircraft noise footprints affect large areas around airports. A control situation must be chosen but it is meaningless to choose a noisy site close to a motorway, for example, for the control!

Community noise has an acknowledged effect on sleep and phase by phase sleep organization. We believe that loss of sleep can have indirect effects on health. Lercher and Kofler[111] in a study of blood pressure and cholesterol levels in people living next to an alpine motorway with heavy truck traffic 24 hours a day, think that the main variable is actually whether windows are closed at night or not. Individual sensitivity is as important in the expression of annoyance (direct factor) as in its effects on health (indirect factor) and explains to some extent the overall variability observed in human responses in research into annoyance and non-auditory effects on health.

Two major surveys carried out in Australia and the Netherlands have not yet been completed (cf. Sydney Congress on Noise Effects 1998).

4.3 Mental diseases and some other noise effects

4.3.1 Psychiatric disorders (cf. Berglund-Lindvall 1993)
Can annoyance due to noise lead to psychological and pathological disorders through time? Epidemiological studies in France show that noise is a recognized cause of depression and that this evolution can be progressive - from annoyance to stress, from stress to sleep disturbance and finally to mild psychiatric disorders - and extremely sudden breakdown. The nature of noise sources - identifiable or annoyance level traffic noise - seems to have an effect on the way annoyance evolves (Vallet[113]).

Stansfeld[114] has observed that in women, depression is linked to sensitivity to noise and that this sensitivity, although decreasing slightly, still remains high after curing. In men, interactions between sensitivity, exposure to noise and psychological morbidity have been observed. Finally, sensitivity to noise permits predicting the psychological reaction to it. This line of thinking suggests the impossibility of filtering or ignoring non-significant sound stimuli in persons with psychopathological tendencies. This theory of psychiatric disorder has been in existence for some time and is pertinent regarding reaction to noise. If persons predisposed to psychiatric illness are less able to ignore non-pertinent stimuli (including sound stimuli), then exposure to noise is likely to have a greater impact on them than on others. This possibility deserves additional research. Moreover, the real or perceived variability of control and predictability requires research on the impact of these variables on theoretically predictable results.

The theory of powerlessness suggests that uncontrollability leads to cognitive modification, loss of motivation, emotional changes and even depression (Maier and Seligman[124]). The theory of anxiety considers that it is the unpredictable nature of noise rather than its uncontrollability which is the decisive factor in the effects observed, including in particular exaggerated fear and anxiety. Exposure and sensitivity to noise have been factors used in forecasting admissions to psychiatric hospitals and depressions in particular (Stansfeld[114]); powerlessness when faced by an event and the resulting anxiety are factors that can be used to explain certain depressions. One way of examining the possible role of uncontrollability/unpredictability and of the subject's powerlessness is to study the responses to certain questions asking people to give their level of agreement on the following assertions:

- "the government does not do enough/take into account noise pollution" and
- "what is the use of complaining, since nobody will do anything anyway".

The second assertion can be considered as an expression of powerlessness. It appears that agreement with the first assertion is more pertinent with respect to noises due to the government (and specifically the Ministry of Defense) than for sources of civilian noise. Applying these theories would permit predicting that with less controllability, there would be more agreement on the expression of powerlessness among populations living close to noisy installations managed by the government (Job[115]).

Regarding depression, a limited survey has been carried out for the Press as has some in-depth work on depression. Results are restricted to a single country to reflect the social and medical characteristics of the healthcare system. The existence of noise-related pathologies was observed in a survey of 350 doctors - general practitioners, specialists and hospital physicians - practicing in France. To the question "Do any of your patients have a noise-related pathology", 30% of all doctors replied yes (42% of general practitioners, 25% of specialists and 12% of hospital physicians). We did not record the percentage of the doctors patients with pathologies of this nature (1994). A full survey (Servant[116]) was conducted with 1.310 doctors all over France into the epidemiology of depressive disorders. The results extracted from this study concern noise from the 3rd analysis phase onwards.

- 1st phase: 27% of all patients suffer from depression.
- 2nd phase: 20% of all patients complain to their doctors about problems

related to environmental nuisances. Depressive pathologies are extremely sensitive to environmental factors - both as a cause of disorders and for their persistence: 4% of all patients think that problems are caused by nuisances while 53% estimate that they play an important role.

- 3rd phase: 63% of the doctors spontaneously ranked noise as the most important factor in stress. This percentage varies from region to region: from 71.4% in Paris to 52.6% in Western France.

This work undoubtedly needs to be confirmed, but it is certainly interesting since it was not only dedicated to noise.

On a purely neurophysiological level, it is blatantly obvious that whatever the results of experiments carried out under the strictest control, working in a noisy environment represents an indisputable biological load. Either the noise is filtered and therefore masked by a central process, which implies increased work by certain structures which, consequently, cannot fulfil other functions, or that the noise is permanently analyzed by the brain. In this hypothesis, this implies time-sharing in processing information directly linked to the task and that directly linked to the noise. What is more, it is well known that it is impossible to give attention at the same time to two pieces of information generated by two different sensorial modes. The limited duration of these experiments and the situations in which they are performed do not correspond in any way to the problems raised during everyday work, where pollution sources are not limited only to the acoustic environment. As we know, these can be perceived differently according to exposure to noise outside the workplace. Lastly, inter-individual differences have not been taken into account in these approaches whose results are far from answering the questions raised.

4.3.2 Other effects

Once again we find ourselves faced by the problem raised by reality: noise, wherever it is generated, and most especially at work, does not occur independently from general conditions. It is often associated with vibrations, ambient temperatures far from thermal neutrality, with a mental load of a challenging nature (Manninen[117]) and, for nearly all the subjects, with stressful physical conditions. On the basis of the studies carried out it is clear that the association of these different factors leads to an exaggeration of reactions, in particular cardiac frequency and arterial pressure. Furthermore, these results have been highlighted by Okada et al.[118] for the association of a continuous noise at 101 dB and vibrations at 5 Hz (acceleration 5.0 m/s) of the entire human body and by Rentzch *et al.*[119] when associating a noise of 80-90 dB and a temperature of 30-40°. This leads us to deduce (Manninen) that the high risk of coronary diseases among Norwegian bus is linked more to the stress of driving than to noise. This association also increases the risk of loss of audition. These results point in the same direction as those obtained from bus drivers in London and locomotive drivers in Russia (Ostapkovich and Perekest[31]) among whom mental stress linked to driving leads to greater loss of audition than that observed among bus conductors and assistants in locomotives.

The following conclusions, according to which different environmental factors are accumulative, are also backed up by the relations demonstrated in 1968 between the loss of audition and depressive syndrome in boiler construction workers. A very strong association between central nervous system problems and loss of audition also exists in coal mining and in the textile industry (Trifonov and Mandaliev[120]).

4.3.3 Noise and vigilance at the wheel

The physical factors affecting car drivers such as noise, vibrations and infrasounds have been studied in simulators and under real conditions.

Studies have also been made on noise sources and levels in driving compartments and their effects on the driver and passengers. The masking of useful sounds signaling faults in vehicle operation (e.g. flat tire) is a potential source of danger, but it is above all the effect of noise on attention that has been analyzed most. The noise level inside vehicles varies as a function of a large number of parameters: model of vehicle, speed, windows open or closed, quality of road surface, etc. Nonetheless, the noise spectrum inside the vehicle is loaded with low frequencies (2 to 63.5 Hz), with levels reaching 100 dB in this part of the spectrum.

Increasing speed leads to a rise of the level centered on the octave bands up to 32 Hz. Opening a window causes an increase of about 15 dB and from 4 to 20 Hz for a vehicle moving at nearly 100 km/h and 8 dB, from 4 to 16 Hz, at 80 km/h. The use of weighting A permits expressing noise levels by a single value but it omits low frequencies. Levels expressed in Leq vary, on a road circuit, from 66 dB(A) for an up-market vehicle with an automatic gearbox to 77 dB(A) for a light, low-powered vehicle.

Fluctuations of noise and radio messages play a not inconsiderable role in maintaining vigilance compatible with driving. Germain[121] observed that heavy-goods-vehicle drivers increased the volume of their radios when feeling tired, in order to stay awake. The effects of noise in a truck cab have been studied by using a simulator, on 17 men with an average age of 34 years, exposed for two hours to 80 dB(A). The main task was driving with central vision, which was not influenced by noise. However, the secondary task of detecting a signal with peripheral vision, was affected by noise since the number of errors increased with noise, though with greater inter-individual variability. On the biological level, electroencephalogram recordings showed decreased vigilance during the two hours of driving, with the appearance of a slower α rhythm (relaxed waking or waking with eyes closed) with noise. Cardiac frequency and its variability increased considerably with noise, which also led to more blinking and peripheral vasomotor reactions. During the two hours, the progressive fall in the tonus of the nape muscles was less discernible in the presence of noise.

Under these experimental conditions, a realistic noise level activated vigilance but it is impossible to know whether this exposure leads to fatigue in the case of prolonged driving, with corresponding effects on vigilance. Research carried out by Fakhar *et al.*[122] showed that a noise level of 80 dB(A) for a journey of four hours on a highway decreased the vigilance of the driver after three hours of driving, after having slightly increased during the first two hours. This positive then negative effect of noise on vigilance was confirmed later (Vallet and Fakhar[123]).

5 Conclusion

Noise represents an indisputable biological load for the organism. On the neurophysiological level it triggers a central filtering process, which implies increased mobilization of certain cerebral structures. This constant stress exerted by noise on control structures, and which is the cause of habituation, has a biological, metabolic and energetic cost that can now be evaluated by brain imaging methods. However, the reactivity of the autonomous nervous system, especially cardiovascular response, does not show any habituation. The effects on the endocrine system, in particular the increase of hormonal secretions, are expressed by non-specific responses characterizing stress.

All these biological mechanisms, whose manifestations are substantially modified by psychological factors, appear very clearly during sleep. After several years' exposure, persons living in noisy places show less reactivity to noise than when the situation is new. However, complete habituation to noise has not been observed, even after long exposure. In the area of disturbance of sleep by noise, results from research are sufficiently numerous and solid to be used for regulatory purposes. The authorities responsible for regulating noise have scientific information in terms of energy and number of events characterized by their maximum levels.

Another way of evaluating the effect of noise on health is by measuring the excess consumption of drugs by populations exposed to noise in comparison with non-exposed groups: more medical consultations, barbiturates and psychotropic drugs, in spite of the low control of important behavioral factors such as wearing earplugs and closing windows at night. Epidemiological surveys on the effects of environmental noise, and no longer on noise at work, are starting to provide results, underlining cardiovascular problems. Nonetheless, these works have difficulty in aggregating all the specific health problems (increased blood pressure, sleeping problems, occurrence of slight depressions, increase of cholesterol rates, absenteeism, etc.) in an overall and single index of the effects of noise on health. This leads to the supposition that the magnitude of these effects has been under-estimated by research which only focuses on a single possible effect of noise instead of on all the effects grouped together.

Research results show statistical trends and significant increases of certain effects in noisy areas, underlining the role of noise in the occurrence of certain pathologies. The presentation of these works in terms of increased frequency of health problems due to noise should start by identifying the frequency of these problems in a population not exposed to noise. On the basis of available results, it is possible to make a rough estimation that noise can be considered as a factor leading to health problems in 5% of the population exposed; however, its role in favoring the occurrence of these problems is far more frequent.

References

[1]   Lambert, J. & Vallet, M. (1994): Study related to the preparation of a communication on a future EC noise policy. Report N° 9420, INRETS.

[2]   Bourdeau, M. & Stanners, M. (1996): *Report to the European Commission*.

[3]   Berglund, B. & Lindvall, T. (1995): *Community Noise: Archives of the Center for Sensory Research*. Vol. 2, Stockholm University and Karolinska Institute.

[4]   Rylander, R. *et al* (1993): *Introduction à la médecine de l'environnement*, ed. Frison-Roche, Paris, 151 pp.

[5]   Laborit, H. (1986): *L'inhibition de l'action*, ed Masson, Paris, 332 pp.

[6]   Jansen, G. (1969): Effects of noise on the physiological state, in *Noise as a Public Health Hazard*. Report ASHA, N° 4, pp. 89-98.

[7]   Muzet, A. (1980): *Modifications végétatives entraînées par le bruit au cours du sommeil*. Rapport CEB-CNRS, Ministry of the Environment N° 76-22.

[8]   Vallet, M., (1991): Night aircraft noise index and sleep research results. *Proceedings InterNoise*, Vol. 1, pp. 207-211.

[9]   Mouret, J. & Vallet, M. (1995): *The Effects of Noise on Health*. Ministry of Health, Paris, published by CIDB, 113 pp.

[10]   Rylander, R., Sorensen, S. & Kajland, A. (1972): Annoyance reaction from aircraft noise exposure. *Journal of Sound and Vibration*, **24**, pp. 419-444.

[11]   Schwarze, S. & Thompson S. (1993): Research on non auditory physiological effects of noise since 1988. Review and perspectives. *Proceedings of Noise and Man*, Nice, Actes INRETS, pp. 252-259.

[12]   Nordic Research Group on noise effects. *Final Report 1997*, 545.

[13]   WHO (1993): *Community Noise. Environmental Health Criteria Document*. External Review Draft, edited by B. Berglund & T. Lindvall, Stockholm, Stockholm University and Karolinska Institute.

[14]   *IEH Report on the Non-auditory Effects of Noise*. Medical Research Council N° R10, 1997.

[15]   Lercher, P., Stansfeld, S. & Thompson, S. (1998): Non-auditory health effects of noise: review of the 1993-1998 period, *Proceedings of Congress on Noise Effects*, Sydney, pp. 213-221.

[16]   Thompson, S. (1997): Cardiovascular and fetal effects of noise, in *IEH Report on the Non-auditory Effects of Noise*, N° R10. Leicester, UK, pp. 70-75.

[17]   Schwela, D. (1998): WHO Guidelines on Commistry Noise. *Noise Effects Congress*. Proceedings, Sydney, pp. 475-480.

[18]   Muzet, A. (1999): *Le bruit*. Flammarion, Paris.

[19]   Porter, N. (1997): Night noise contours: a feasibility study. *NPL Report*, London.

[20]   Hyde, I.H. & Scalapino, W. (1918): The influence of music upon electrocardiograms and blood pressure. *American J. Physiol.* **46**, pp. 35-38.

[21]   Ward, D. (1993): Developments in noise-induced hearing loss during the last 25 years. *Actes congrès Noise and Man*, Vol. 3, Nice.

[22]   Loth, D., Menguy, C. & Teyssou, M. (1994): Effet sur la santé de l'écoute de la musique à haut niveau sonore. *Convention Ministère des affaires sociales, de la santé et de la ville*, 338 pp.

[23]   Selye, H. (1962): *Le stress de la vie*. Gallimard, Paris.

[24]   Jansen, G. (1998): Health concepts and noise effects. *Proceedings of Congress on Noise Effects*, Sydney, pp. 697-702.

[25]   Young, J.S., Upchurch, M.B., Kaufman, M.J. & Fechter, L.D. (1987): Carbon monoxide exposure potentiates high-frequency auditory threshold shifts induced by noise. *Hearing Research*, **26**, pp. 37-43.

[26]   Fechter, L.D., Young, J.S. & Carlisle L. (1988): Potentialisation of noise induced threshold shifts and hair cell loss by carbon monoxide. *Hearing Research*, **34**, pp. 39-48.

[27]   Sullivan, M.J., Rarey, K.E. & Conolly, R.B. (1989): Ototoxicity of toluence in rats. *Neurotox, & Teratol.* **10**, pp. 525-530.

[28]   Muijser, H., Hoogendij, E.M.G. & Hooisma, J. (1988): The effects of occupational exposure to styrene on high-frequency hearing threshold. *Toxicology*, **49**, pp. 331-340.

[29]   Rebert, see in Mouret & Vallet [ref 9].

[30]   Crofton, see in Mouret & Vallet [ref 9].

[31]   Ostapkovitch, V.E. & Perekrest A.I. (1969): Concerning the hazard criteria of the sound analyzer in the workers of locomotive brigades. *Otorinolaryngology*, **2**, pp.25-29

[32]   Garcia, A.M. (1999): Los niveles de prevencion de los riesgos laborales. *Gac. Sanit.* **13(3)**. pp.173-176.

[33]   Rabinowitz, J. (1999): Substances ototoxiques et effets du bruit à l'hôpital. *Med. Hyg.* **57(1)**, pp.600-601.

[34]   Rechtschaffen, A. & Kales, A. (1968): *A Manual of Standardized Terminology Techniques and Scoring System for Sleep Stages of Human Subjects*. U.S. Government Printing Office, Washington D.C.

[35]   Borsky, P. (1962): *Community Relations to Sonic Booms in the Oklahoma City Area*. Aerospace Med. Res. Lab., National Opinion Research Centre.

[36]   Lambert, J., Simonnet, F. & Vallet, M. (1984): Patterns of behaviour in dwellings exposed to road traffic noise. *Journal of Sound and Vibration*, **92(2)**, pp. 159-172.

[37]   Langdon, F.J. (1976): Noise nuisance caused by road traffic in residential areas. *Journal of Sound and Vibration*, **47**, pp. 243-282.

[38]   Page, M.A. (1976): Une enquête longitudinale sur la perturbation du sommeil par le bruit routier. *Recherche environnement*, **3**, pp. 11-43.

[39]   François, J. (1979): *Les répercussions du bruit des avions sur l'équilibre des riverains d'aéroports. Etude longitudinale autour de Roissy - 3ème phase*. Rapport IFOP BV 77.333 pour le Ministère de l'Environnement. Paris.

[40]   Fidell, S. & Jones, G. (1975): Effects of cessation of late night flights on an airport community. *J. Sound Vibr.* **42**, pp. 422-437.

[41] Ollerhead, J. *et al.*, (1992): *Report of a Field Study of Aircraft Noise and Sleep Disturbance*. UK Department of Transport, London Civil Aviation Authority.

[42] Fidell, S., *et al.* (1995): Field study of noise induced sleep disturbance. *J. Acoust. Soc. Am*, **98**, pp. 1025-1033.

[43] Griefahn, B. (1975): Effects of sonic booms on fingerpulse amplitudes during sleep. *Arch. Occup. Environ. Hlth.* **36**, pp. 57-66.

[44] Lukas, J.S. (1972): Awakening effects of simulated sonic booms and aircraft noise on men. *J. Sound Vibr.* **20**, pp. 457-466.

[45] Muzet, A. & Erhardt, J. (1978): Amplitude des modifications cardio-vasculaires provoquées par le bruit au cours du sommeil. *Cœur et médecine interne*, **17**, pp. 49-56.

[46] Globus, G., Friedman, J. & Cohen H. (1973): Effects of aircraft noise on sleep recorded in the home. *Sleep Res.* **2**, pp. 116-122.

[47] Vallet, M., Gagneux, J.M., Blanchet, V., Favre, B. & Labiale, G. (1983): Long term sleep disturbance due to traffic noise. *J. Sound Vibr.* **90**, pp. 173-191.

[48] Wilkinson, R. (1984): Disturbance of sleep by noise: Individual differences. *J. Sound Vibr.* **1**, pp. 55-63.

[49] Griefahn, B. & Gros, E. (1983): Disturbances of sleep. Interaction between noise, personnel and psychological variables. *Noise as a Public Health Problem*, Edizioni Tecniche Amplifon, Milano, pp. 895-904.

[50] Ehrenstein, W. & Muller-Limmroth, W. (1978): Laboratory investigations into the effects of noise on human sleep. *Proceedings Congress Noise as a Public Health Problem*, Repport N° 10, pp. 433-441.

[51] Vallet, M., Blanchet, V., Bruyere, J.C. & Thalabard, J.C. (1977): La perturbation du sommeil par le bruit de circulation routière: étude in situ. *Collection Recherche Environnement, Ministère de l'Environnement*, Paris, N° 3, pp. 183-212.

[52] Ohrström, E. & Bjorkman, M. (1983): Sleep disturbance before and after traffic noise attenuation in an apartment building. *J. Acoust. Soc. Am.* **73**, pp. 877-879.

[53] Metz, B. & Muzet, A. (1977): Effets propres et interaction de l'élévation du niveau sonore et de la température ambiante sur le someil. *L.D.F., Recherche et Environnement*, **3**, pp. 81-160.

[54] Friedmann, J. & Globus, G. (1974): Effects of cessation of late night landing noise on sleep electrophysiology in the home. NASA Case, Report 132543, Washington D.C.

[55] Blois, R., Debilly, G. & Mouret, J. (1978): Daytime noise and its subsequent sleep effects. *Noise as a Public Health Problem. Proceedings of the 3rd International Congress*, Rockville, Tobias Rep. N° 10, pp. 425-432.

[56] Lukas, J.S. (1975): Noise and sleep: a literature review and proposed criteria for assessing effects. *J. Acoust. Soc. Am.* **58**, pp. 1232-1242.

[57] Ouis, D. (1999): A review on noise and sleep. *Noise Health*, NRN, University College London, pp.11-36.

[58]   Williams, H. (1973): Effects of noise on sleep: a review. *Proc. International Congress on Noise as a Public Health Problem*, pp. 501-511.

[59]   Berry, B. & Thiessen, G.J. (1970): *The Effect of Impulsive Noise on Sleep*. Nat. Res. Council of Canada. Publ. NRC 11.

[60]   Collins, W.E. & Iampietro, P.F. (1972): Simulated sonic booms and sleep: effects of repeated booms of 1.0 psf., FAA, Office Aviation Medicine, Rep. OAM-72.35.

[61]   Ludlow, J.E. & Morgan, P.A (1972): Behavioral awakening and subjective reactions to indoor sonic booms. *J. Sound Vibr.* **25**, pp. 479-495.

[62]   Lukas, J.S. (1972): Awakening effects of simulated sonic booms and aircraft noise on men. *J. Sound Vibr.* **20**, pp. 457-466.

[63]   Muzet, A., Scheiber, J.P., Olivier-Martin, N., Erhardt, J. & Metz, B. (1973): Relationship between subjective and physiological assessments of noise-disturbed sleep. *Proc. of the 2nd Congress on Noise as a Public Health Problem*, pp. 575-585.

[64]   Thiessen, G.J. (1978): Disturbance of sleep by noise. *J. Acoust. Soc. Am.* **64**, pp. 216-222.

[65]   Griefahn, B., Jansen, G. & Klosterkotter, W. (1976): *Zur problematik lämbedingter Schlafstörungen. Eine Auswertung von Schlaf-literature.* Umwelt Bundes amt Berlin.

[66]   Osada, Y., Ogawa, S., Ohkubo, C. & Miyzaki, K. (1974): Experimental study on sleep interference by train noise. *Bull. Inst. Publ. Health*, **23**.

[67]   Rice, C. (1972): *Sonic Boom Exposure Effects: Sleep Effects*. Inst. Sound and Vibr. Res., Southampton.

[68]   Vernet, M. (1979): Effects of train noise on sleep for people living in houses bordering the railway line. *J. Sound Vibr.* **3**, pp. 656-674.

[69]   Griefahn, B. (1977): Long term exposure to noise. Aspects of adaptation, habituation and compensation. *Waking and Sleeping*, **1**, pp. 383-386.

[70]   Fidell, S., Horonjeff, R., Teffetteller, S. & Pearsons, K. (1982): Behavior awakening as functions of duration and detectability of noise intrusions in the home. *J. Sound Vibr.* **84**, pp. 327-336.

[71]   Pearsons, K. (1996): Recent field studies in the United States involving the disturbance of sleep from aircraft noise. *Proc. Inter-Noise '96*, **5**, pp. 2271-2276.

[72]   Latham, D.: Personal Communication.

[73]   Frusthorfer, B. (1983): Daytime noise stress and subsequent night sleep: interference with sleep patterns, endocrine function and serotoninergic systems. *Noise as a Public Health Problem, Turin*, **44**, pp. 1015-1018.

[74]   Carter, N. (1996): Transportation noise, sleep and possible after effects. *Environment International USA*.

[75]   Carter, N., Ingham, P., Tran, K. & Hunyor, S. (1997): A field study of the effects of traffic noise on heart rate and cardiac atrhythmia during sleep *J. Sound Vibr.* **169(2)**, pp. 211-227.

[76]   Griefahn, B., *et al.* (1998): What nighttimes are adequate to prevent noise

effects on sleep. *Congress Effects of Noise Proceedings*, Sydney, pp. 445-450.

[77]   Carter, N. (1998): Cardiovascular response to environmental noise during sleep. *Congress Effects of Noise Proceedings*, Sydney, pp. 439-444.

[78]   Maschke, C., Harder, J., Hecht, K. & Balzer, H., (1998): Nocturnal aircraft noise and adaptation. *Proceedings Euronoise Congress*, Munich, pp. 253-259.

[79]   Kawada, T. & Suzuki, S. (1999): Change in rapid eye movement (REM) sleep in response to exposure to all night noise and transient noise. *Arch. Env. Health*. **54.**4, pp. 336-340.

[80]   Ando, Y. & Hattori, H. (1973): Effects of intense noise during fetal life upon postnatal adaptability (statistical study of the reactions of babies to aircraft noise). *J. Acoust. Soc. Amer.* **27**, pp. 101-110.

[81]   Vallet, M. & François, J. (1982): Evaluation physiologique et psychosociologique de l'effet du bruit d'avion sur le sommeil. *Travail humain* **45**, pp. 155-168.

[82]   Wehrli, B. & Wanner, H.U. (1978): Auswirkugen des Strassenverkehrsläms in der Nacht. *Kampf dem Lärm*. **25**, pp. 138-149.

[83]   Relster, J. (1981): *Effects of Traffic Noise on Psychical Health*. National Danish Road Lab.

[84]   Vallet, M., Cohen, J.M., Mosnier, A. & Trucy D. (1999): Airport noise and epidemiological study of health effects: a feasibility study. *InterNoise Proceedings*, Fort Lauderdale, pp. 1327-1330.

[85]   Lamure, C. (1982): Le problème de l'indice de bruit nocturne pour les riverains de voies de circulation. *Revue d'Acoustique* **62**, pp. 180-186.

[86]   Large, J. (1981): Airport Noise Standards. *New Scientist* **12**, pp. 651.

[87]   Maurin, M., Lamure, C. *et al.*, (1982): Le problème de l'indice de bruit nocturne pour les riverains de voies de circulation. *Revue d'acoustique N°62*, pp. 180-186.

[88]   Ollerhead, J. B. (1978): Variation community responses to aircraft noise with time of day. *Noise Control Engineering* 11, pp. 68-78.

[89]   Rice, C. (1980): Time of day corrections to aircraft noise metrics. FAA/NASA Workshop NASA CP 2135/FAA-EE-80-3.

[90]   Brooker, P. (1980): Civil Aviation Authority - DORA Report 8008. *Aircraft Noise and Sleep Disturbance: Final Report*, London.

[91]   Diamond, I.A. & Walker, J.G. (1986): An interim study of the influence of residual noise on community disturbance due to aircraft noise. *InterNoise '86*, pp. 941-946.

[92]   Schultz, T.J. (1978): Synthesis of social surveys on noise annoyance. *J. Acoust. Soc. Amer.* **64**, pp. 377-405.

[93]   Rucker, A. (1975): Strassenverkehrslärm in Wolhgebieten. *Kampf dem Lärm* **22**, pp. 72-81.

[94]   Vallet, M., Letisserand, D., Olivier, D., Laurens, J.F. & Clairet, J.M. (1988): Effects of road traffic noise on the rate of heart beat during sleep. *Proc. 5th Int. Congr. on Noise as a Public Health Problem*, Stockholm.

[95]   De Jong, R., Knottnerust, T. & Altena, K. (1986): Nachtvluchte en hun

invloed op slaap. Gezondheiden gedrag. Rap. TNO N° 86011, Leiden, 50 pp.

[96]   Tullen, J., Kumar, A. & Jurriens, A. (1986): Psychophysiological acoustics of indoor sound due to traffic noise during sleep. *J. Sound Vibr.* **110(1)**, pp. 129-141.

[97]   Eberhardt, J.L. (1988): Health effects of community noise. Nordic council of ministers. Nordic Noise Group. 45 pp.

[98]   Rabinowitz, J. (1991): Les effets physiologiques du bruit. *La Recherche* **22**, N° 229, pp. 178-187.

[99]   Pearsons, K., Barber, D. & Tabachnick, T. (1989): Analyses of the predictability of the noise induced sleep disturbance. HSDTR89029 for US Aircraft NSBIT.

[100]   Griefahn, B. (1992): Noise control during the night, *Acoust. Austral.* **20**, pp. 43-47.

[101]   Kuwano, S., Namba, S. & Mizumani, N. (1998): The effects of noise on the effort to sleep. *InterNoise '98*. Christchurch, pp. 1083-87.

[102]   Rosenman, R.H. & Friedman, M. (1977): Modifying type A behavior pattern. *J. Psychosom. Res.* **21**, pp. 323-332.

[103]   Zhao, Y.M., Zhang, S.Z., Spear, S. & Spear R.C. (1991): *British Journal Ind. Medicine* **4**, pp. 179-184.

[104]   Lang, T., Fouriaud, C. & Jacquinet-Salord, F. (1992): *International Archive of Occupational Environmental Health* **63**, pp. 369-372.

[105]   Fogari, R., Zoppi, A., Vanasia, A., Marasi, G. & Villa, G. (1994): *Hypertension Journal* **12**, pp. 475-479.

[106]   Talbott, E. *et al.* (1996): Occupational noise exposure, use of hearing protectors over time and the risk of high blood pressure: the results of a case-control study. *Proceedings InterNoise '96*, Vol. 4, pp. 2131-2136.

[107]   Hirai, A., Takata, M., Mikawa, M., Yasumoto, K., Lida, H., Sasayama, S. & Kafamimori, S. (1991): *Hypertension Journal* **9**, pp. 1069-1073.

[108]   Hessel, P.A. & Sluis-Cremer, G.K. (1994): Occupational noise exposure and blood pressure: longitudinal and cross-sectional observations in a group of underground miners. *Archives of Environmental Health* **49**, pp.128-134.

[109]   Kristal-Boneh, E., Melamed, S., Harari, G. & Green, M.S. (1995): Acute and chronic effects of noise exposure on blood pressure and heart rate among industrial employees: the Cordis study. *Archives of Environmental Health* **50**, pp. 298-304.

[110]   Thompson, S. (1996): Noise auditory health effects of noise: updated review – *Proceedings InterNoise '96*, Vol. 4, pp. 2177-2182.

[111]   Lercher, P. & Kofler, W. (1993): Noise as a public health problem, INRETS, *Proc. Noise and Man '93*, Vallet, M. (ed.), France, Vol. 3, pp. 465-468.

[112]   Babisch, W., Ising, H., Gallacher, J., Sweetnam, P. & Elwood, P. (1999): Traffic noise and cardiovascular risk: the caerphilly and speewell studies, third phase, 10 year follow up. *Arch. Env. Health* **54.3**, pp. 210-216.

[113]   Vallet, M. (1995): Annoyance, sleep disturbance, stress and psychiatric

disorders. *15th ICA Trondheim Proceedings*, pp. 319-322.

[114] Standfeld, S.A. (1992): Noise, noise sensitivity and psychiatric disorder: epidemiological and psychophysiological studies. *Psychol. Med. Monograph Suppl.*, 22.

[115] Job, S. (1993): Psychological factors of community reactions to noise. *Proc. Congress Noise and Man*, INRETS N° 34, Vol. 3, pp. 48-59.

[116] Servant, J. (1993): Anxiété et troubles anxiodépressifs liés au bruit – Thèses, Dissertation, Université de Lille, France.

[117] Manninen, O. (1985): Cardiovascular changes and hearing threshold shifts in men under complex exposure to noise, whole body vibration, temperatures and psychic load. Tampere, Finland

[118] Okada, A., Kajikawa Y. & Nohara, S. (1984): Combined effects of climate and noise. *Combined Effects of Environmental Factors*, Manninen, O. (ed.), Keskupaino.

[119] Rentzch, M., Prescher, W. & Weinreich, W. (1984): The combined effects of climate and noise on labour efficiency and stress. *Combined Effects of Environmental Factors*, Manninen, O. (ed.), Keskupaino.

[120] Trifonov, H. & Mandaliev, O. (1973): Issledovanie vlijanija suma na sluhori apparat artefilanoe davlenic i nervruja sistema rabocih tekstilnoi promyslennostic. *Scient. Congr. Industrial Hygiene Noise, Dust and Occupational Diseases*, Varna.

[121] Germain C. (1989): Le métier de chauffeur routier, gestionnaire de contraintes. Thesis Dissertation, Lyon-France.

[122] Fakhar, S. *et al*, (1991): La posture corporelle comme indicateur. *Le maintien de la vigilance dans les transports*, Vallet, M. (ed.). Editions Paradigme, pp. 151-158.

[123] Vallet, M. & Fakhar, S. (1999): Effects of noise and vibration on the vigilance of drivers of light vehicles, accepted for publication in *J. Vehicle Design*.

[124] Maier, S.F. & Seligman, M.E.P. (1976): Learned helplessness: theory and evidence. *Journal of Experimental Psychology: General*, 105, pp. 3-46.

Chapter 4

Community response to environmental noise

Rainer Guski
Faculty for Psychology, Ruhr-University Bochum, Germany

1 What is community noise and community response to noise?

1.1 Community noise

The term community noise denotes a major part of environmental noise which impinges on many residents. It comprises mainly traffic noise from road, air, rail, and ship transportation, in addition, noise from construction sites, factories, workshops, and leisure activities belong to this class insofar as residents are affected. Although the individual community noise sources may be very specific, they share some characteristics: (1) they are public in the sense that the sound immission is not confined to areas owned by the source administrators, (2) they are chronic rather than acute, (3) they consist of a series of single events with a certain variation over time, and (4) individual respondents rarely can do anything about them.

Although specific data are rather scarce, the European Commission [3] estimated that about 20% of the European population are exposed to daytime noise levels equal or above $L_{Aeq} = 65$ dB(A) - a level that is seen as unacceptable by many scientists. Fig. 1 gives some of the results of a comparison by Lambert & Vallet [1], showing the percentage of residents of some European countries exposed to day-time road traffic noise equal or above 65 dB(A) L_{Aeq}. It should be noted that these data are not fully comparable, because (1) in Spain, all noise sources are included, and (2) the definition of the day varies between countries. Still, it is evident that a great proportion of the Spanish population is exposed to very unhealthy noise levels, and that the northern parts of Europe seem to be less exposed to road traffic noise than the middle and southern parts.

It should be noted that about 50% of the mileage of European private cars is due to activities not connected to work, education, or household activities. This does not mean that half of the road traffic noise is due to unnecessary leisure activities, but it does mean that a considerable portion of road traffic noise is due to activities almost totally under the control of individual drivers. They can choose the distance of leisure activities from their home, the means of transportation, the route, the speed, and the time to travel. Usually, these choices are made with respect to ease and fun, not with respect to their impact on the environment, or the health of other people.

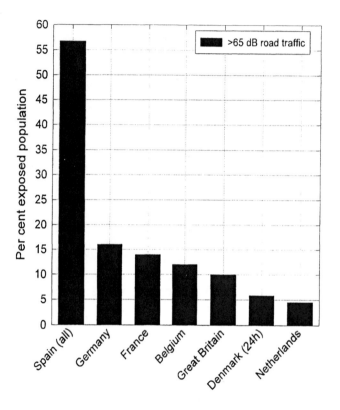

Fig. 1: Estimated per cent of residents exposed to daytime road traffic noise above 65 dB(A) L_{Aeq} (after Lambert & Vallet [1], Umweltbundesamt [2]). The exact definition of daytime varies between countries.

More than 40% of the European population are estimated to live in so-called gray areas, i.e. areas exposed to day noise levels between L_{Aeq} = 55-65 dB(A). For most communities in industrial countries, the main noise problem is road traffic noise, because (1) there are many more people exposed to road traffic noise than to any other public source, and (2) in general, the level of road traffic noise tends to increase over the years. The European Commission [3] estimated 90% of the European population exposed to day noise levels above 65 dB(A) to suffer from road traffic noise, 1.7% from aircraft noise, and 1% from railroad noise at the same high level.

The European Commission [3] also estimated the proportion of residents exposed to traffic noise of L_{Aeq} 65 dB(A) during daytime to have been constant over the past 15 years, while the proportion of residents in gray areas increased; in addition, the nighttime levels of road traffic noise increased significantly. With aircraft noise, the proportion of people exposed to high L_{Aeq}-levels is said to have decreased over the past 20 years, but the duration of quiet periods also decreased dramatically. Furthermore, the authorities of international airports tend to allow more and more air traffic at the border times of a day (early morning and late evening). Today, residents at airports complain that they are exposed to never- ending aircraft operations

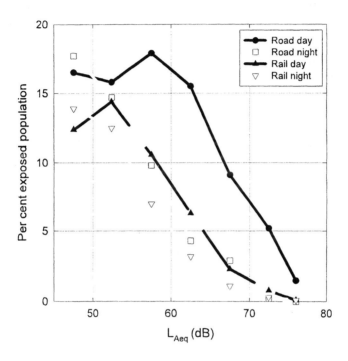

Fig. 2: Per cent of German residents exposed to road and rail noise in 1992 (Umweltbundesamt [2]).

Some countries provide somewhat more precise estimations: For instance, about 31% of the German population was exposed to daytime (06.00-22.00h) road traffic noise above 60 dB L_{Aeq} in 1992 and 1997, about 7.5% was exposed to nighttime road traffic noise at the same levels, while 9.5% was exposed to daytime railroad noise, and 4.7% to nighttime railroad noise at the same levels in 1992 [2]. About 16% of the German population is exposed to daytime road traffic noise above 65 dB L_{Aeq}. Both for railroad and air traffic noise, about 1% of the German population is estimated to suffer from daytime levels of 65 dB L_{Aeq}. Unfortunately, the same percentage is estimated for nighttime railroad noise. Due to the increasing volume of road traffic, as well as to the increased traveling speed, the level of road traffic noise has increased in Germany: Ullrich [4] estimated an increase of about $L_{Aeq} = 2$ dB(A) at German highways in the past 20 years. At residential streets, he estimated an increase of about 1.5 dB.

1.2 Community response to noise

Taken literally, the term "community response" denotes activities of a community caused by noise. The community consists of groups of individual residents of a circumscribed local area, and organizations (or institutions) responsible for the administration. The literal translation of "community response to noise" therefore would mean the regulations and actions taken by groups and administrators against noise [5]. Seen this way, the smallest units of community response to noise are complaints of individual members of a community, and the greatest units are regulations of the administration against noise. Citizen organizations and pressure groups are somewhere in the middle range of community units.

This chapter will deal with low-level units of community responses to noise, i.e., at the level of individual members of a community who took part in systematic field studies. In most of these studies, residential areas have been chosen by acoustical criteria, comprising a single noise source at a certain range of levels, and residents of these areas have been chosen by sociodemographic criteria. The data are mostly gained by means of personal face-to-face, telephone, or written interviews. Very rarely, medical investigations have been undertaken in addition to the interviews, and the reactions of vulnerable groups (children, sick, and old people) are still less investigated. The average responses of residents at certain levels are called community response to noise.

2 Community responses

2.1 Effects on public health

Epidemiologic studies try to assess the prevalence of diseases in high noise areas, compared to the prevalence at low noise areas. They usually need large samples in order to get a significant statistical effect. It is almost impossible to prove the cause of a sickness in individual cases, but it is possible to estimate the statistical risk of getting a disease by living in an area exposed to high levels of noise in contrast to living in an area less exposed. Although the results of these studies are not conclusive in every case, there is sufficient evidence to maintain that road traffic and aircraft noise at high levels increase the risk of cardiovascular disease.

2.1.1 Road traffic noise effects

Eiff *et al.* [6] compared 192 young residents of a traffic noise area ($L_{Aeq} > 63$ dB(A)), and a control area ($L_{Aeq} < 55$ dB(A)), who had recently moved into the respective areas. For one and a half years, these persons were tested every six months for blood pressure and different chemical, psychophysical, and social variables. At the end, no long-term influence of traffic noise could be ascertained, but the vascular reactivity to test noise was increased in the residents of the noisy area.

In a large study in the United Kingdom, about 4800 masculine long-term residents exposed to road traffic noise, and aged between 45 and 59 years, were subjected to tests of biological risk factors for ischemic heart disease [7]. It was not possible to establish a clear dose-response relation between calculated noise level and one of the risk factors, but a statistical contrast analysis between the lowest and the highest exposure groups (51-55 dB vs. 66-70 dB L_{Aeq}) showed a slight increase of hemostatic blood lipids, and the authors suggest an increase of the expected relative risk (1.1) for ischemic heart disease (IHD), especially in men aged 45-60 years. In a follow-up study 10 years later, a relative risk of IHD incidence of 1.2 was found in the highest noise group (only one of the two initial cohorts was available).

The same authors used 243 and 4035 men for hospital- and population-based case-control studies of the relative risk of myocardial infarction. The levels of the respective road traffic noise at the subjects' homes were calculated, ranging between 60 and 80 dB L_{Aeq}. It was not possible to show a clear dose-response function, but the relative risk of a myocardial infarction increased at the highest level (71-80 dB) of road traffic noise among men in the order of 1.2 to 1.3 [7]. Unfortunately, no independent research group has tried to replicate these results up to the present.

2.1.2 Aircraft noise effects

In one of the earliest studies, Eiff *et al.* [8] made extensive medical examinations on 392 residents of the old Munich airport and found increased systolic and diastolic blood pressure with increasing levels of aircraft noise. This finding was supported by Knipschild [9] at the Amsterdam airport. In addition, Knipschild & Oudshoorn [10] found an increase of medication purchase with increasing nighttime aircraft noise.

A study on school children [11] compared the blood pressure in pupils attending schools underneath the flight paths of Los Angeles International Airport (inside average L_{Amax} = 74 dB) with matched controls in less exposed schools (56 dB). It turned out that the blood pressure of the children living under the flight paths was significantly higher than that of controls, especially in children who were exposed less than 2 years. The effect could not be established in a longitudinal study [12], probably because of systematic mortality effects: The families of children with high blood pressure were more likely to move out of the high noise areas. (For a review, see also Morrell *et al.* [13].)

In order to test the influence of persistent daytime aircraft noise on the excretion of cortisol (one of the so-called stress hormones), Maschke *et al.* [14] compared the cortisol excretion of 28 residents (aged 35-65 years) living near Tegel Airport in Berlin (a) in an experimental situation involving nighttime aircraft noise, and (b) with a control group of residents living in a quiet neighborhood. Tegel Airport has no significant nighttime air traffic. The nighttime cortisol excretion was extrapolated over 24 hours, and it turned out that the average excretion of the airport group was significantly higher than that of the control group. In addition, the airport group was experimentally subjected to recorded aircraft noise (16-64 flights at $L_{Amax} = 55/65$ dB) during the night, and the data suggest that the cortisol excretion was further increased after disturbed nights.

The opening of the new Munich Airport in 1991 provided an opportunity to assess long-term and short-term effects of aircraft noise on pupils [15]. A total of 217 pupils (mean age 9.9 years) from different rural areas were tested for blood pressure, and different components of urine, each 6 months before the opening of the new airport, and 6 and 18 months afterwards. The noise areas were exposed to 53 dB L_{Aeq} (24 h) before, and 62 dB after the opening - with L_{01} being 73 dB(A). The control areas were exposed to 53 dB L_{Aeq} (24 h) before, and 55 dB afterwards - with L_{01} being 64 dB(A). It turned out that each of the physiological variables increased after the opening of the airport not only in the noise areas but in the control areas as well, but the increase of systolic blood pressure and in norepinephrine was greater in noise areas. The authors conclude that the aircraft noise did increase the stress of the children significantly. In addition, life quality judgments of the children decreased systematically after opening of the airport in the noise groups.

It is difficult to evaluate the data from the rather short publication. For instance, it is not revealed how many children were in the respective groups, how reliable the acoustic data are, and the potential effect of statistical regression is not discussed. The fact that 5 of the 6 variables before noise exposure are lower in the noise group than in the control group is not explained, and the increase of the noise group means may partially be seen as a statistical regression effect: the means of the noise group are underestimated in the first measurements, and repeating the measurements reduces the measurement error. Still, this statistical explanation cannot account for all of the variance between groups, since it is improbable that a regression effect occurs with 5 of the 6 variables at the same time.

Cardiovascular effects have been tested with respect to military low-flying aircraft (MLAF), the results being inconclusive. On the one hand, Ising *et al.* [16] found increased blood pressure levels in girls from a highly exposed area (125 dB L_{Amax}; 65 dB L_{Aeq}) shortly after individual simulated overflights, compared to girls from a less exposed area (112 dB L_{Amax}; 59 dB L_{Aeq}). On the other hand, no significant effect occurred in boys, nor any in a field study with real overflights. (For psychosomatic and psychiatric effects, see also Poustka & Schmeck [17].)

In a review of military aircraft noise and health, Flindell [93] points to many methodological problems associated with studying the effects of irregular, loud and infrequent noise events. In addition, "acoustic startle effects and associated physiological responses can equally be interpreted either as signs of healthy reactivity or as precursors of disease. MLAF events can contribute to acute annoyance at the time that they occur, but the possible mechanisms for noise-related annoyance to contribute to longer term morbidity through some vague and unspecified stress hypothesis are obscure" (p. 351).

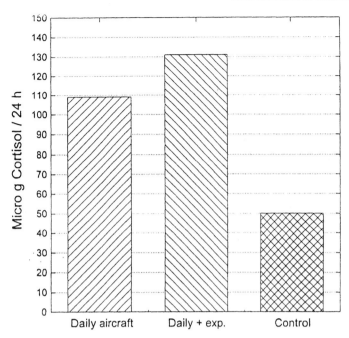

Fig. 3: Average 24 h cortisol excretion for residents near an airport (left two columns) and a control group (right column). The center column represents cortisol excretion after noisy nights (after [14]).

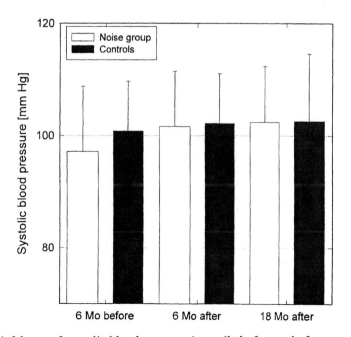

Fig. 4: Means of systolic blood pressure in pupils before and after opening of an airport (Evans *et al.* [15]). The thin vertical lines are standard deviations.

2.2 The annoyance response

The major effect of environmental noise on a community is annoyance, i.e. the likelihood of annoyance responses in a community study increases significantly with the number and level of environmental noise events. This increase is generally more systematic than any other reaction to noise, and many politicians or judges consider the number of highly annoyed residents to be the most useful indicator of a community noise problem.

The most widely quoted definition of annoyance was formulated by Lindvall & Radford [19]: Annoyance is a "feeling of displeasure associated with any agent or condition known or believed by an individual or a group to be adversely affecting them". But a recent study sheds some doubt upon this rather emotional concept: Noise experts from 8 nations scaled the annoyance response to be highly associated with both disturbance, and nuisance [18]. That is, the annoyance response is seen as comprising both a behavioral response (disturbance) and a negative evaluation: The residents are disturbed by noise, and at the same time, they are somewhat angry about the circumstance that they cannot do much against a situation they don't want. Therefore, annoyance judgments of residents exposed to community noise for years cannot be easily compared with annoyance judgments (or loudness judgments) of subjects in short-term laboratory studies.

The annoyance response considered here is mainly the response of residents participating in systematic questionnaire studies, using one or more questions about the general evaluation of living conditions with respect to noise annoyance. In the past, the annoyance questions varied between studies. For instance, Langdon [20, 21, 22] used the question: "Looking at this card, would you tell me which number represents best how you feel about noise around here?", and the response format was a 7-point numeric scale labeled "satisfactory/unsatisfactory" at the lower/upper end points of the scale. In contrast, Tracor Inc. [23] asked respondents to rate the disturbance of 9 intended activities (e.g., relaxing outside, sleeping, conversation, telephone) on a 5-point scale, and one of the annoyance responses (Annoyance G) was the sum of the 9 disturbance ratings. Another annoyance response (Annoyance V) was a weighted sum of direct annoyance ratings related to different situational aspects of the neighborhood. Other studies use a direct verbal rating of annoyance, similar to the one used by Björkman *et al.* [24]: They asked respondents to rate the annoyance from different sources in their respective environment, and presented a verbal scale containing 4 points (from "not at all" to "very").

In future, many studies will probably follow the recommendations proposed by Team No. 6 of the International Commission on the Biological Effects of Noise (ICBEN), see Fields *et al.* [25]: They recommend using at least two common general annoyance questions: One of them uses an 11-point numeric scale, the other uses a verbal scale. Both questions do the following: explicitly identify the particular noise source, directly rate that noise source, explicitly request negative reactions, only focus on general reactions rather than a particular disturbance, and immediately ask all respondents about the full range of annoyance (from none to the highest degree) rather than screening on the basis of a dichotomous question. The question should be: "Thinking about the last (...12 months or so...), when you are here at home, how much does the noise from (...noise source...) bother, disturb, or annoy you?". After completing an 10-nation scaling study (Felscher-Suhr *et al.* [26]), the team determined that the verbal scale to be used in future studies should consist of 5 steps, which in English are labeled "extremely, very, moderately, slightly, not at all".

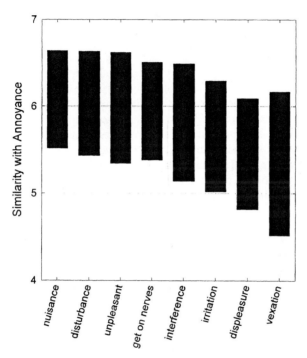

Fig. 5: Medians and interquartil ranges of similarity judgments of noise experts with respect to annoyance (Guski *et al.* [18]).

Often, the number (or percentage) of highly annoyed people is calculated from continuous response scales. Schultz [27] introduced the convention to count the number of people choosing the top 27%-29% of a response scale on annoyance as indicating the number of "highly annoyed" (HA) people. For instance, if the response scale has seven steps, with "1" indicating the lowest, and "7" the highest degree of annoyance, this convention would count the top 2 or 3 steps (6/7, or 5/6/7) to show "high annoyance"; on a 5-point scale, the top two steps would be taken, and on an 11-point scale, the top 3 points. Although there is no empirical evidence that it is valid at all, this procedure is widely adopted for supporting political decisions, and for comparing different studies. Following this tradition, Oliva [28] reports that 25% of residents exposed to road traffic, or aircraft noise at a day-level of 60 dB L_{Aeq} are highly annoyed; Gjestland *et al.* [29] report 25% HA at day-levels of 56,5 dB L_{Aeq} aircraft noise, and Fields & Walker [59] report 25% "bothered" at 24 h-levels of 58 dB L_{Aeq} railroad noise. In comparing different sources of noise at the same evening noise level (18.00-22.00 h) of 65 dB L_{Aeq}, Finke *et al.* [31] found 70% HA with aircraft noise, 59% HA with industrial noise, 55% HA with road traffic noise, and about 45% HA with railroad noise.

2.3 Reported disturbances

Residents exposed to environmental noise primarily complain about a series of disturbances: Disturbance of communication activities outside and inside the house (telephoning, listening to radio or TV, speaking with friends etc.), disturbances of the night (falling asleep, early awakening), disturbance of recreational activities, and disturbance of concentration (e.g., during work), to name but the most prominent disturbances. For instance, at daytime levels of 65 dB L_{Aeq}, about 72% of the residents were disturbed while speaking outside the house, 55% inside the house, 49% while listening to radio/TV/music, and 28% during recreational activities in a study on aircraft noise effects [29]. In a subgroup of residents exposed to road traffic noise (mean daytime level of 70.6 dB L_{Aeq}), Finke *et al.* [31] found 63% residents highly disturbed in activities outside the house, 45% in communication, 42% in recreational activities, 38% in activities inside the house, and 36% in sleep.

Communication disturbance is most likely to occur with peaks in noise level (e.g., during aircraft overflights, or a passing by of trucks and trains), because the environmental sounds mask the sounds intended to pick up. During active communication, they force the speakers to raise their voice, which both puts some strain on the vocal tract, and adversely affects the setting for private talking (cf. Lazarus [32]). These disturbances are very likely to be remembered, and the self-rating of communication disturbance usually is a very reliable answer in systematic interviews which shows high correlations with different noise descriptors, especially with maximum noise levels. The absolute ratings of communication disturbance outside the house is usually larger than that inside the house.

Disturbances during the night are reported by many residents exposed to road, railroad and aircraft noise with frequent noise events. The reported disturbances comprise difficulties falling asleep, awakenings during the night, an early end of the night, and a low quality of nighttime recreation. (For a detailed questionnaire of nighttime disturbances, see Olivier-Martin *et al.* [33]). The correlations with acoustic variables describing the energy (e.g. L_{Aeq}) usually are significant, but on a lower level, compared to reported communication disturbances. With road traffic noise, certain acoustic variables describing background noise levels (e.g., L_{90}) or "quiet periods" during the night [34], sometimes show a greater association with reported disturbances of the night than do energy-related or maximum levels [35]. It is difficult to interpret reported nighttime disturbances, because (a) the statistical associations between reported sleep quality variables and physiological indicators of sleep quality generally are low [35], and (b) reported sleep quality variables often show good statistical associations with daytime noise levels in field studies [36]. The latter effect may be due to the circumstance that the actual feelings and behavior during sleep are not accessible to consciousness, and if residents are asked about disturbances during the night, they may be inclined to make inferences from their day-time disturbances.

Disturbance of recreational activities usually comprises of disturbance of daytime rest periods, hobbies, and using the garden, or balcony, belonging to the residents' property. These disturbances usually are very annoying, because the residents claim their home as their primary territory which is mainly used for recreational purposes, which should only be controlled by themselves, and must not be intruded upon by actions of anybody else.

Disturbance of mental concentration (e.g., during work) is somewhat less prominent as a community response to noise, because there are only a few people working at home. However, if such disturbances occur, they increase the mental effort necessary to accomplish the task, because task-irrelevant sounds attract attention, and detract attention from the task in hand. The performance quality is not necessarily less than in a quiet setting, but the time and effort used for the same task is usually longer, or greater, as shown by laboratory studies (e.g. [37, 38]).

2.4 Effects on children

Children have only recently become subjects of community noise studies. One early focus of research concentrated on effects of aircraft noise on the development of newborns, showing growth inhibition effects of noise on fetuses (cf. [39]), and increased systolic and diastolic blood pressure in children attending primary schools under a flight path [40]. More recent studies show conflicting results with respect to mental performance: While one group of researchers got decreased long-term memory performance with chronic aircraft noise, and no effect on reading performance, another group obtained contradictory results. Classroom experiments with children aged 12-14 years showed that long-term aircraft noise exposure at the old Munich airport before its closing down (about 66 dB L_{Aeq}, compared with control groups, 53-61 dB L_{Aeq}) is systematically associated with decreased long-term text recall. This decrease was lost two years after closing down, while one year after the opening of the new airport, a performance decrease could not be seen [41, 42]. The other group of researchers tested similar issues at Heathrow Airport: Children attending four schools exposed to high levels of aircraft noise (>66 dB L_{Aeq}) were compared with children attending four matched control schools exposed to lower levels of aircraft noise (<57 dB L_{Aeq}). The tests were repeated after one year. At baseline and follow-up chronic exposure to aircraft noise was significantly associated with impaired reading comprehension and higher noise annoyance, adjusting for age, deprivation and main language spoken at home. Chronic exposure to aircraft noise was significantly associated with impaired sustained attention and higher self-reported stress after adjusting for age, deprivation and main language spoken at follow-up. Contrary to previous findings, chronic exposure to aircraft noise did not affect long-term memory and motivation [43].

In areas of low-altitude military flights, parents often fear health effects on small children due to the fast-rising levels of jet aircraft noise. Psychosomatic and psychological variables were studied in children (aged 4-16 years) from two German residential areas, one of them being exposed to low-altitude military flights, the other in the vicinity of a small civil airport. With the same L_{Aeq} level, the subjective disturbance caused by military low-altitude flight noise was essentially greater than that due to ordinary flight noise. A comparison revealed that frequencies of ear symptoms (tinnitus lasting more than one hour and permanent hearing threshold shifts of greater than 30 dB) were higher only in areas where maximal flight noise levels considerably exceeded 115 dB (A) accompanied by rapid noise level increases. Blood pressure measurements yielded significantly higher values (group difference 9 mm Hg systolic) in girls living in these highly exposed areas. They also showed somewhat greater anxiety levels [17, 16].

Very few studies have dealt with the effects of chronic exposure to road traffic noise on children. At present, results point to increased systolic and diastolic blood pressure in preschool children chronically exposed to traffic noise above 60 dB L_{Aeq} (24 h, [44]), and decreased concentration performance of children exposed to daytime traffic noise levels about 69.6 dB L_{Aeq}, compared to children exposed to daytime noise levels about 54.2 dB L_{Aeq} [45]. The authors believe that chronic traffic noise has an adverse effect on attention and concentration performance, especially in situations of increased nighttime traffic. In addition, a study on 796 children aged 8-12 years, living in 13 alpine areas of Austria [46] showed that children chronically exposed to road traffic noise and their concurrent exhaust fumes tend to spend less time outside the house.

2.5 Economic effects

Economic effects of noise include social and individual expenditures on noise protection, and the costs of damage.

The social **expenditures on noise protection** have been estimated for road transportation noise to be about 0.3% of the GNP in OECD countries (Lambert *et al.* [102]). About 92% of these costs are spent for road traffic noise, and 8% for railroad noise. At present, there are no data showing individual expenditures on noise protection (e.g., for sound insulating windows).

The social costs of community noise damage include productivity losses, health care costs, effects on property values, and costs associated with loss of psychological wellbeing.

Productivity losses can occur due to errors made during work, extra-time necessary to complete a task correctly, or fatigue because of disturbed sleep or rest outside work. It is very difficult to estimate the monetary value of such losses, and the respective economic models can be criticized. In Germany the productivity losses due to noise from many sources was estimated to be 0.2% of the GNP [47]. An estimation combining productivity losses and health care costs to remedy physiological and psychological effects of noise for all noise sources together runs to about 17 thousand million EURO for Germany [48].

The effects of noise on **property values** have been studied in many countries, and a summary has been given by Lambert & Vallet [1]. They conclude (a) that the decrease in housing values due to noise varies because of cultural differences and standards of living, but (b) increased from 1970 until now. While the rate of depreciation was negligible or near zero in the 1960s, it increased in the 1970s to approximately 0.3-0.8% per decibel (in the range between 61 and 69 dB L_{Aeq}), and was estimated at 1% per decibel in the middle of the 1980s.

2.6 Effects of noise changes

In former times, predictions concerning the effects of noise changes were rather simplistic: It was believed that respondents in changing situations would react like respondents in stationary situations. Meanwhile, predictions are now rather cautious, because (a) it seems that respondents today are somewhat more annoyed at similar noise levels than respondents some years ago, (b) a short-term decrease of less than 6 dB(A) in noise level does not have significant effects on the number of highly annoyed residents, while a short-term increase of very few dB(A) in noise level does increase noise effects significantly, and (c) even the expectation of a negative change - without any physical change - may increase noise effects. These statements will be backed up by a few examples:

(a) Airport authorities like to support their demand for increasing the traffic volume by stating that the noise level of individual aircraft decreased in the past few years, and at some places, the overall aircraft noise levels (expressed in L_{eq}, or D_{NL}) did really decrease. On the other hand, the number of aircraft flights increased, and the number of low-noise intervals decreased. Generally, residents in the vicinity of airports do not seem to notice small decrements of noise levels [49], sometimes they even react more strongly to similar noise levels today than they did before [50]. This sheds some doubt on the belief that dose-response relationships which have been established 10 years ago are still valid for predicting annoyance in the future. In addition, the percentage of fairly annoyed people in the general public tends to increase over time, while the percentage of highly annoyed persons rests at a constant level, at least in The Netherlands and in Germany.

(b) Empirical studies on the effects of any change of noise levels or noise composition on residents are rare, and they are difficult to perform, because retrospective annoyance judgments are subject to memory bias [51], and the expectations of residents about the effects of change are often not realistic [52]. In addition, their answers may tend to "exaggerate" because of interview-demand characteristics [53]. In a review of the effects of noise level changes,

Vallet [49] showed that level decrements which are less than 6 dB(A) will not be reflected in the percentage of highly annoyed. Raw & Griffiths [54] show that Mean Annoyance (the mean of raw annoyance scores) may be significantly changed by level changes of the order of 3 dB(A), and the effect of changes may be visible up to 9 years after the change [55].

(c) Recently, a study at Sydney International Airport [56] showed differences in self-reported physiological noise reactions to be dependent on noise expectations of the residents. Before any actual change in the flight paths of the airport had been installed, residents expecting a positive change (less noise) reported less physiological symptoms than residents not expecting a change, while residents expecting a negative change (more noise) reported more symptoms. This again points to the limited value of merely looking at dose-response relationships; it also underlines that noise effects are moderated by variables which are not reflected in acoustic measurements.

3 Causal linkage between exposure and annoyance

3.1 Relations between energy and annoyance or disturbance

3.1.1 Dose-response relationships

The relationship between acoustic descriptors and original annoyance or disturbance responses (on a continuous scale) is usually linear and can be expressed by means of correlation coefficients. The amount of covariation depends on a number of factors, e.g., the degree of data aggregation, the range of sound levels, and the homogeneity of sound sources, to name but a few. For instance, a high level of data aggregation diminishes variance and thus leads to higher correlation coefficients than those obtained on an individual data set: while individual-data coefficients between L_{Aeq} and annoyance judgments seldom exceed the order of 0.50, grouped-data coefficients of the same origin may reach the order of 0.90. A small range of sound levels usually leads to low coefficients, and mixed sources in the same data set do the same.

One example of a dose-response relationship between the L_{Aeq} of aircraft noise and a global annoyance/disturbance rating on an 11-point numeric scale is shown in Fig. 6. The data stem from a Swiss study ("Lärmstudie 90", [28, 57]) at Zürich and Geneva, comprising 2052 personal interviews in 58 residential areas. The linear correlation between L_{Aeq} and annoyance judgment (original scale) is r = 0.49 for ungrouped data, and 0.98 for grouped data; the linear correlation between L_{Aeq} and percent highly annoyed (Top 3 of the original scale) is r = 0.35 for ungrouped data, and 0.97 for grouped data. The nonlinear correlation (as shown in Fig. 6) is r = 0.98 for grouped data. Similar correlation coefficients are found with road traffic, or railroad noise, as long as there is just one noise source in a data set. The degree of covariation between the acoustic descriptor and individual annoyance or disturbance responses (ungrouped data) is about 25% in the example shown, and it seldomly exceeds the order of 30% in a field study. That is, the major part of the individual annoyance or disturbance variance cannot be explained by means of an acoustic variable alone. It does not depend significantly on the noise descriptor, as long as two conditions are met: (a) the acoustic variable describes mainly maximum energy in the time frame the response refers to, and (b) the response variable describes annoyance or disturbance. That is, with aircraft, road, and railroad noise, similar correlation coefficients are obtained using L_{Aeq}, maximum level, or number of loud events. Furthermore, the addition of another acoustic variable usually does not increase the correlation with annoyance, because different acoustic variables usually are highly interrelated in a data set on a single noise source. That is, with high L_{Aeq} values, we usually also get high maximum levels and high numbers of loud events. The situation may change, if there is more than one source in a data set, e.g., aircraft and road traffic noise. In this case, the overall correlation between acoustic and annoyance variables decreases, because the level of annoyance differs between different noise sources at the same noise level (see section 3.3), and the correlation between different acoustic variables decreases also.

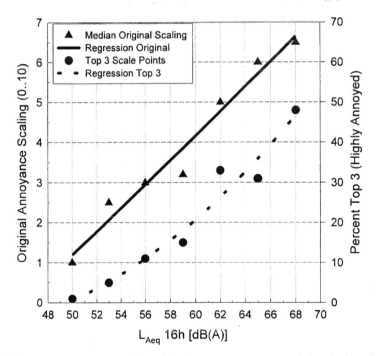

Fig. 6: Dose-response curves for original annoyance scores (triangles) and per cent highly annoyed people (points) in a Swiss aircraft noise study [57].

Another example of a dose-response relationship is shown in Fig. 7. This time, the acoustic descriptor is the logarithm of the number of motor vehicles per peak hour, and the response variable is the average scaling of activity disturbances inside and outside the house due to road traffic noise. Data are from a subsample of Finke *et al.* ([31], N=413). The points show mean values for 13 residential areas with road traffic as the most prominent noise source. The individual disturbance scores are equally well related to L_{Aeq} r = 0.58) as to the logarithm of the traffic volume r = 0.60).

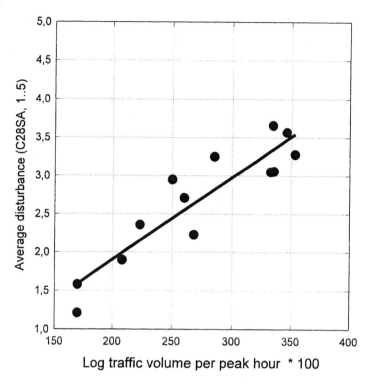

Fig. 7: Relation between road traffic volume and average activity disturbance [31].

3.2 Relations between other acoustic parameters and noise effects

It has already been said that the degree of covariation between annoyance/disturbance variables and acoustic variables usually does not depend very much on the exact definition of the acoustic descriptor, as long as there is only one noise source, and the acoustic variable mainly describes maximum energy. The reason is the high systematic covariation between different acoustic descriptors. For instance, road traffic noise with high daytime L_{Aeq} levels show high L_1 levels, high numbers of loud events, and high nighttime L_{Aeq} levels also. However, it should be noted that annoyance and disturbance of activities are just the major class of community reactions to noise, and they provide no exhaustive description of noise effects. For instance, the satisfaction with recreational aspects of the residential area is heavily reduced by community noise, but this time, maximum energy aspects of the acoustic situation are less efficient in predicting the satisfaction as are acoustic variables describing the base level of noise (e.g., L_{90}, cf. [58]), the level of the least noisy hour during the day, or the number and duration of quiet periods. This has been shown by Finke *et al.* [31] in a subsample of 412 residents living in 12 residential areas exposed to either road traffic noise (6 areas) or one of 6 areas exposed to either aircraft, railroad, or workshop noise. The original satisfaction variable did correlate significantly with daytime L_{Aeq} (r = -0.53), but it correlated somewhat better with the level of the least noisy hour during the day (r = -0.60), and still better with a variable weighting low level periods according to their quietness (PP8D, r = 0.62). The authors conclude that disturbances and general annoyance may be somewhat more closely related to acoustic variables describing maximum energy, while satisfaction variables may be more closely related to acoustic variables describing quietness aspects of the noise situation.

3.3 Source-dependent dose-response relationships

Several European studies have shown that railway noise is less annoying than road traffic noise, at least at levels above 65 dB L_{Aeq} [59, 60, 61]. Most comparisons have concentrated on general annoyance and concluded that both sources have rather similar effects at low levels (about 50 dB L_{Aeq}) - the level difference for equal response varying between 0 and 10 dB, while the effects differ at high levels (about 70 dB L_{Aeq}): the same general annoyance is obtained with road traffic noise level being between 5 and 9 dB less than railroad traffic noise. This observation has led many European administrations to produce a so-called "railway bonus", i.e., the criterion for railway noise abatement should be 5 dB less than that necessary for road traffic noise abatement. However, it should be noted that the so-called "bonus" is greater for certain effects (e.g., reported sleep disturbances), and there is actually a "penalty" for communication disturbances - see Fig. 8. A closer look at special effects of noise reminds one of the fact that both the energy-related measurement of noise as well as the global annoyance reaction are very crude descriptions of dose-effect relationships.

There have been much less studies comparing aircraft and road traffic noise. In the summary given by Fields & Walker [59], three Heathrow aircraft surveys and two on English road traffic noise were compared. The authors suggest that at the noise levels at which many noise regulations are set (above 65 dB L_{Aeq}) similar general annoyance levels are met with aircraft noise being 5-10 dB lower than road traffic noise. In contrast, Oliva [28, 57] reports a very close correspondence between dose-response relationships for aircraft and road traffic noise at two Swiss airports in the range between 50-68 dB daytime L_{Aeq}.

The reasons for different dose-effect relationships with respect to general annoyance are only partially understood: It may well be that the energy-type of description for different sources of noise is too crude in order to reflect different qualities of the sound environment -- for instance, road traffic noise at high energy levels usually has a rather continuous character, while railroad noise usually has high peak levels, and a number of quiet periods in between. This may point to a high weighting of quiet periods in the formation of general annoyance, but if this was the main reason for preferring railroad noise over road traffic noise, the same reason could not hold for preferring road traffic noise over aircraft noise. On the other hand, it is assumed that moderating variables are different for different noise sources - for instance, residents evaluate railroad traffic to be less dangerous than road traffic noise [62], and road traffic noise to be less dangerous than aircraft noise [23, 63]. This also may only partially explain different dose-effect relationships, because a recent study [61] has shown that positive evaluations of railroad traffic do not effectively moderate annoyance and disturbance reactions.

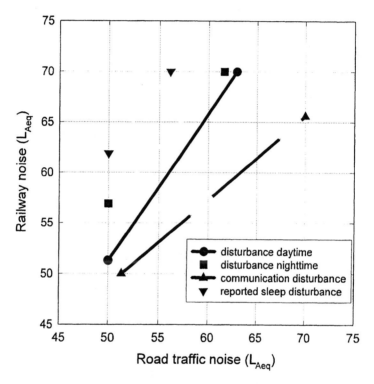

Fig. 8: Comparison between road traffic and railroad noise levels producing the same disturbance effects (after Schuemer & Schuemer-Kohrs [60]).

3.4 Effects of source combinations

In some residential areas, people are exposed to more than one noise source. Generally, road traffic noise is the major noise source, but in some areas, noise from aircraft, railroad, or industrial sources add to the general noise climate. For planning purposes, it has often been found necessary to know how residents feel in such situations, and it seems reasonable to assume that residents are able to provide both a separate annoyance judgment for each of the sources as well as a total annoyance judgment about the noise source combinations. It was expected that the total annoyance would be somewhat like a sum of the two specific annoyances, or it should be at least the maximum of either of the two specific annoyances. But field studies had puzzling results: The total annoyance was less than the maximum of either of the specific annoyances [64,65, 66]. In the study of Miedema & Van den Berg [67], 31% of the respondents reported a total annoyance that was less than the maximum of either of the specific annoyances. For a recent review, see Ronnebaum *et al.* [68].

Since there is no general explanation for this result, Guski [69] questioned some of the assumptions held in models of the relation between source specific and total annoyance. These assumptions are:

1. The specific noise sources build a common auditory stream which residents process as one form. One alternative would be that they process separate auditory streams for each source. This is probable in situations which combine acoustically very different sources, e.g., motorway noise and tramway noise. Another alternative would be that residents process temporal combinations of noisy events as separate auditory streams, like the railroad crossing alarm together with the train passing by [70].

2. The specific noise sources have comparable effects which will combine easily to a common effect. The alternative would be that specific sources have specific effects. For instance, if the main effect of railroad noise is disturbance of communication, and the main effect of road traffic noise is disturbance of recreation, it is improbable that residents exposed to both sources will easily respond to a question of total effects and total annoyance.

3. The total annoyance judgment evaluates disturbances and other noise effects in the same way as the specific annoyance judgments do. The alternative would be that specific disturbances are evaluated differently, depending on the source. For instance, it is probable that the disturbance of communication is evaluated as less severe than the disturbance of recreation.

4. Judgments of total annoyance comprise all situations relevant for specific noise effects. The alternative, as proposed by Diamond & Rice [65] would be that residents use different temporal frames of reference for different sources.

5. The total annoyance is moderated by the same situational and personal variables as the specific annoyances do. The alternative would be that different moderators are effective for different sources, as well as for total annoyance. For instance, if the annoyance to road traffic noise is moderated mainly by the personal noise sensitivity of the residents, and the annoyance to aircraft noise is moderated mainly by fear of aircraft crashing in the neighborhood, we will not be able to predict their respective moderating functions for total annoyance. (For a discussion of moderating variables, see section 4.)

In reviewing results from former field studies, Guski [69] concluded that in noise source combinations that include road traffic noise, (1) both the reactions to road traffic noise and the reactions to the other source are strongly influenced by the non-road traffic noise level, (2) the total noise evaluation is also strongly influenced by the non-road traffic noise level - but see [71], and (3) the effects of noise in residential areas exposed to multiple sources depend more on moderating variables (such as the individual coping capacity of the residents) than on acoustic descriptors of noise.

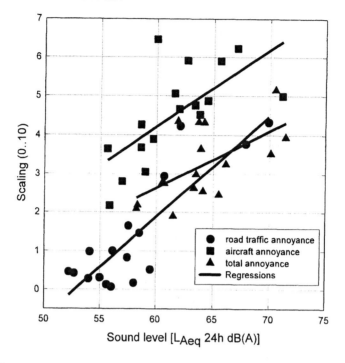

Fig. 9: Dose-response relations for source specific and total noise annoyance in the study of Taylor [64]. Please note that the total annoyance is less than the maximum specific annoyance at comparable levels.

4 Moderators of annoyance

It has been shown in the preceding sections of this paper that there are statistically significant dose-response relationships between acoustic parameters as descriptors of the noise situation in a residential area and certain noise effects, such as annoyance judgments. It has also been shown that the systematic covariation between such data sets very rarely exceeds 30%. Since a further 30% of the reaction variance often can be predicted by variables intervening between acoustic load and reaction, we have to deal with such intervening variables which usually are called moderating variables, or for short: moderators.

On a theoretical level, Baron & Kenny [72] distinguish between moderators and mediators. Moderators are variables which are not affected by the environmental agent, but they covary with the reaction variable - that is: moderating and raction variables may depend on each other. Mediators can be seen as "primary reactions", they depend on the stimulus variable, and they also influence the "secondary reaction". See Fig. 10 for an illustration. For instance, if a person's age influences the degree of annoyance expressed in an interview, age is a potential moderator, because it is sure that the age variable does not depend on the noise level at the person's home. On the other hand, if noise affects the health status of a person, and this in turn affects the annoyance judgment, the health status is a potential mediator. We will restrict our discussion to moderators, because they are much more prominent than mediators in social psychological field studies.

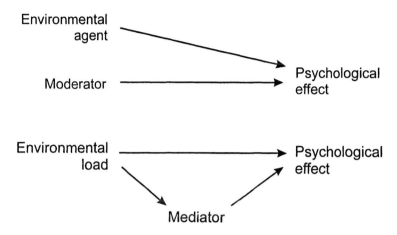

Fig. 10: The theoretical distinction between moderating and mediating variables in the relation between environmental load and psychological effect.

4.1 Time of day, weekend and seasonal effects

It is obvious that noise events occurring during the night have more serious effects with respect to annoyance and disturbance of the residents than the same events have during the day. This difference is partially caused by the respective vulnerability of different activities during night and day, and in addition, by different noise (or quietness) expectations of the residents during different parts of the day. Data from different field studies show that residents emphasize the need for quiet nights and quiet evenings [e.g. 31, 73, 74]. During the night, residents do not want to be disturbed while sleeping, and during the evening they do not want to be disturbed while in the family circle, or while being engaged in recreational activities. From a large US survey, Fields [73] concluded that the day can be roughly divided into four noise sensitivity periods consisting of two relatively steady periods, night (24.00-05.00 hours) and day (09.00-16.00 h), and the early morning and evening transition periods. Maschke *et al.* [75] argue that due to circadian biological rhythms, humans are more sensitive to distracting stimuli during the transition period between night and day.

Weekends differ from weekdays in that the morning transition period is 1 h later, and the numbers of people engaged in aural communication during the day at home are approximately 1/2 or 3/4 greater [73]. In addition, many national administrations restrict noisy activities on Sundays and holidays, and this in turn has led to the expectation of tranquility on these days.

In comparing the times of greatest disturbance mentioned by residents, Finke *et al.* [31] showed that residents generally describe the early and late evening hours (15.00-22.00 h) to be disturbed most, followed by the early morning hours (06.00-08.00 h). With respect to the evening hours, there was a remarkable crossover of disturbed hours between road traffic and aircraft noise: while the disturbance due to road traffic noise decreased from the early evening hours until the night, the disturbance due to aircraft noise increased and had its maximum in the late evening hours. This seems to be characteristic for international airports following a nighttime traffic restriction, since the overseas flights tend to arrive as late as possible.

There are also seasonal effects, especially with disturbances of outdoor activities, and with respect to general annoyance, too. For instance, when representative samples of the German population are asked about the annoyance due to road traffic noise during the summer period (May-September), the percentage of annoyed people is about 70, even if the summer period is not mentioned in the question. If they are asked the same question during the winter period, the responses is about 10% less [76].

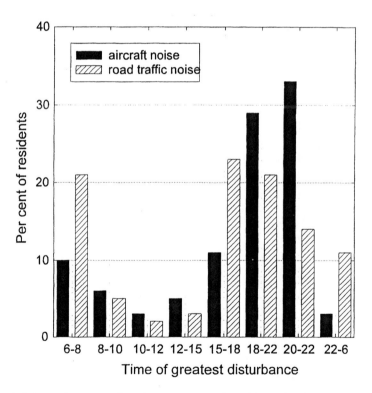

Fig. 11: Time of greatest disturbance, as mentioned by subgroups of residents in the study of Finke *et al.* [31]. Please note that the airport did not have nighttime traffic.

4.2 Social characteristics

In this chapter, I like to distinguish between social and personal characteristics as moderators of noise reactions. Personal characteristics are variables which are linked tightly to an individual resident, show a considerable stability over time and situations, and vary between individuals considerably. Social characteristics are linked to noise situations and are shared to a considerable degree between individuals of a society. For instance, age, sex and socio-economic variables (such as education or income) are linked to individuals, they are considered as personal characteristics. On the other hand, the social evaluation of a noise source, as well as the social evaluation of noise abatement authorities, are considered as social characteristics. For instance, in our society, railway systems are generally evaluated more positively than airlines, because railways are perceived as being less dangerous to the general public. In addition, the publication of a recent aircraft crash is a social factor, and will influence the evaluation of all airlines to a considerable degree.

Personal characteristics are thought to be the result of the individual development of a person, while social characteristics are thought to be the result of social developments, shared by larger groups of the society. Of course, there is no clear-cut distinction between personal and social characteristics, because the individual development usually takes place within a certain society.

Personal and social characteristics may vary within a group of residents in the vicinity of a noise source, but they are not dependent on the level of noise. On the other hand, personal and social characteristics depend on each other, and mostly we find personal variance within social characteristics. But there is at least one reason for discussing them separately: While personal characteristics cannot be used for noise abatement programs, social characteristics can. For instance, if residents mistrust source authorities, they will not trust their respective noise abatement programs, and noise annoyance will stay high. But if they trust independent authorities and see that these are controlling the noise abatement program, the same program will be much more effective in reducing the annoyance.

We will discuss four social factors: (1) the general evaluation of the noise source, (2) trust or misfeasance with the source authorities, (3) the history of noise exposure, and (4) expectations of the residents.

4.2.1 General evaluation of a noise source
Most citizens have an "image" of different noise sources, even if they have never lived near one of them. For instance, railroad traffic has the image of an old, reliable, and rather harmless means of transportation that is under the control of a public institution. In contrast, road traffic is seen as dangerous and under private control, and air transportation is also seen as dangerous and under the control of an institution that makes money and tries to gain as much money as possible. On the other hand, the acoustic features of these three noise sources are different. Therefore, it is not easy to tell why aircraft noise annoyance is usually greater than road traffic noise annoyance, and railroad noise annoyance is usually less than road traffic annoyance at the same energy levels. But a field study

comparing 7 different noise sources [31] proposed that the general (non-acoustic) evaluation of a noise source contributes much to the noise annoyance. For instance, the evaluation of a source as being unhealthy was the most important non-acoustic predictor for all important annoyance reactions, and its standardized regression coefficient proved to be higher than that of the energy-based exposure variable in three of the five tests.

In several other field studies, the "importance of the source" was used as a potential moderator. For instance, it was believed that residents near an airport, who accept the economic importance of air transportation, and hope to have a direct economic benefit from it in the long run, would show less annoyance than residents who do not believe in the airport importance. Typically, the correlations between "importance" and annoyance judgments were in the order of –0.15 to –0.28, but the moderating effect was rather low.

4.2.2 Trust or misfeasance with source authorities

The concept of trust or misfeasance goes back to Borsky [77]. He called the degree to which residents felt that the source agents do everything in order to avoid unnecessary noise "consideratedness". Similar terms have been used by McKennell ([78] "preventability"), and Schuemer [79], "trust in the good will of responsible people"). This chapter uses the term "misfeasance", as proposed by Tracor Inc. [23], because it denotes the current situation at several airports and railroad tracks at least in Germany. The residents know that they cannot control the noise emissions, they fear that the noise source and/or the noise itself does some harm, and they usually assume that an institution supposed to be under public control is not really controlled by the public, and it certainly does not give priority to public health. This attitude contributes to noise annoyance. For instance, the residents at Düsseldorf Airport are well known in Germany, because they usually produce higher annoyance scores than residents at other comparable airports. This is not only due to a very active group that opposes the unrestricted increase of the air traffic, but also due to the behavior of the county administration, which used to own the airport and recently sold it to a private company which spent some million German Marks more in order to get permission for a further increase in aircraft movements of the order of 30%.

The degree of misfeasance is often measured on a scale containing items concerned with "the people who run the airlines", "the airport officials", "the government officials", "the pilots", the designers and makers of airplanes", and "the community leaders" (Leonard & Borsky [80]). The summated ratings show correlations with annoyance variables of the order of 0.25-0.36, and regression analysis showed "misfeasance" to contribute to annoyance to a significant degree, even after controlling for "fear" and "health attitudes". See Table 1 for results of an (uncontrolled stepwise) Multiple Regression Analysis of Leonard & Borsky [80]. This analysis can be questioned with respect to methodological reasons, but the importance of misfeasance can clearly be seen - it is well above the importance of the acoustic variable CNR. This variable will increasingly play a major role in environmental psychology.

Misfeasance attitudes by residents might likely be reduced by actions of the noise authorities who show (a) clear data about the acoustic situation and its development, (b) an acceptance of the existence of harmful effects of noise, (c) clear data about noise abatement programs, and (d) a willingness to communicate and cooperate with the residents.

Table 1: Multiple regression analysis.

Criterion: June-July Annoyance. $R = 0.76$, $R^2 = 0.58$

Predictor	r	β
fear	0.72	0.50
health attitude	0.63	0.28
misfeasance	0.32	0.08
CNR	0.32	0.04

Leonard & Borsky [80]. (CNR = Composite Noise Rating.)

4.2.3 History of noise exposure

In land use planning, the history of the land use is often seen to be important. For instance, if a road transportation company wants to settle in a residential area, it will not be allowed to do so in Germany, because it does not fit in to the history of the area. If the company is already located in the vicinity of a residential area and wants to expand in the direction of the residents, it will be much easier to get permission, because transportation noise is already there. On the other hand, many residents live about 10 to 20 years in the same area in Germany. They tend to say that their living area has become louder in the past few years and that the amount of noise does not accord any more with the original character of the area. With aircraft noise, this impression stands in sharp contrast to the observation of acousticians who state that, in general, the energy-based measures of global exposure have decreased. It seems that the residents do not react to global energy, but instead to noisy events, i.e., to the amount, distribution, duration, levels and the meaning of acoustic immissions. Therefore, it is not surprising that residents near an airport say the noise has grown louder. They probably mean that the number of aircraft has increased, and the duration of relatively calm intervals decreased. In addition, the increase in the number of aircraft movements does not conform to the history of the area. A similar idea has been expressed by Fidell [81], suggesting that the prediction of noise annoyance should take into account the historical distribution of noise levels. Unfortunately, we don't have any empirical test of this idea yet, but if it is true, then plans for new noise sources, or expanding the noise immission area to residential areas which have not been exposed before, should be considered with great caution.

4.2.4 Expectations of residents

This aspect has much to do with the historical position, but also with subjective control and misfeasance: residents in the vicinity of motorways, long-distance railroads, and airports have experienced noise increases during the past years. Very often, they fear that this increase will go on, and they don't see effective means forthcoming for stopping this "progress". The German Noise Abatement Society (DAL, personal communication, see also [82]) has often found this aspect to be the major source of motivation for residents' complaints who live near airports and railroads. These people are disturbed by noise, but they are also annoyed because they expect an increase in the amount of disturbances without having any influence on the development. There are no meaningful data on this topic, but the noise annoyance will probably decrease if the residents know that the future level of noise exposure (not only the level of sound energy) and its distribution in their living areas will at least not increase any further.

4.3 Personal characteristics

Sometimes so-called "personality traits" were considered to moderate the influence of noise, e.g., the well-known concepts of "extraversion vs. introversion", or "neuroticism". In a similar vein, Schick [83] reflected on laboratory studies showing the huge intra-individual variance in sound evaluation, and the general influence of personal styles of perception on the completion of distraction tasks. However, none of the general personality traits proved to be useful in explaining the variance of noise annoyance judgments in field studies. Instead, there are few personal variables which are more specifically related to environmental exposures but are still not reaction variables (which would covary with the degree of environmental exposure in a cross-sectional study). We will consider (1) sensitivity to noise, (2) fear of harm connected with the source, (3) evaluation of the source, and (4) capacity to cope with noise.

4.3.1 Sensitivity to noise

In one of the earliest major studies on aircraft noise annoyance, McKennell [78] used a single question for assessing the sensitivity of the respondents to noise: "Would you say you were more sensitive or less sensitive than other people are to noise?", and the respondents had four options: "more sensitive/less sensitive/same/don't know". This question was put after a long set of questions concerning aircraft noise, and we do not know whether the respondent's reactions to this question covary with noise exposure or not. Anyway, McKennell [78] showed that the self-reported degree of sensitivity to noise has two specific effects on annoyance judgments: (1) At the same level of noise exposure, sensitive persons generally show significantly more annoyance than non-sensitive persons. (2) The covariation between noise exposure level and annoyance is rather shallow in persons at both extreme ends of the sensitivity scale, whilst the covariation is steep for persons lying in the middle of this scale (cf. Fig. 12).

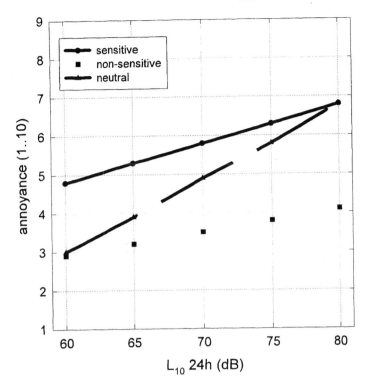

Fig. 12: Influence of noise sensitivity on noise annoyance in the study of McKennell [78].

In several other studies, noise sensitivity has been defined more explicitly and measured with somewhat greater care. For instance, most of the German field studies I know of use a set of items describing individual feelings in noisy situations, and the respondents are asked to rate the agreement with each item (Likert format). If possible, these items should be presented before discussing effects of noise in a residential area. Here are examples of the items proposed by Guski *et al.* [84]: "I am disturbed, if doors bang all the time", "I can't stand listening to several radio or television sets at the same time", or "I get nervous if a dog barks all the time". Noise sensitivity is then calculated as the average of all respective items.

A prominent sensitivity scale has been published by Weinstein [85] and uses 21 items in a Likert format, e.g., "I am more aware of noise than I used to be", "At movies, whispering and crinkling candy wrappers disturbs me". Unfortunately, the whole set of items is too long in order to be used in big samples, but a subset of nine items will probably do, as has been shown with a German translation of the Weinstein scale by Zimmer & Ellermeier [86].

In comparing seven old field studies of aircraft noise, Schuemer [79] considered the sensitivity (or susceptibility) to noise often to be the third best predictor of aircraft noise annoyance - after exposure and fear of aircraft crashing. While the correlations between individual noise exposure level and individual annoyance were in the order of 0.25-0.68, the correlations between sensitivity and annoyance were in the order of 0.23-0.28. The reviews of Job [63] and Fields [87] come to similar conclusions: noise sensitivity usually shows significant covariation with annoyance judgments, the order of correlation coefficients varying between 0.15 and 0.48. Some sensitivity scales tend to be influenced by the level of environmental noise and thus may reflect a reaction; this should be avoided in order not to strain the statistical assumptions of the moderator model.

4.3.2 Fear of harm connected with the source

Several aircraft noise studies have shown that the (subjective) fear of an aircraft crashing in the neighborhood is related to noise annoyance. Also, the general belief that the noise source - not the noise itself - has some detrimental effect on health, often shows significant correlations with non-aircraft noise. From my knowledge, the respective history starts again with McKennell [78, 88]. He asked two questions related to aircraft effects: "Would you say that the aircraft have any effect on your own health?" (Q.18a), and "Do you think they have any effect on other people's health?" (Q.18b). It seems to be quite typical that many people deny harmful effects for themselves, but agree with the second question, assuming harmful effects for other people. When using the answers to the second question as a moderator, it turned out that those residents who believed in harmful source effects showed significantly greater noise annoyance than those who did not believe it. Moreover, the covariation between level and average annoyance was significantly greater for the latter group (see Fig. 4). It should be remembered that this aspect does not ask for the potential harmful effects of noise, but instead for the harmful effects of the source.

In his early review, Schuemer [79] noted the fear variable to be the most important non-acoustical variable to be mentioned in aircraft noise studies, followed by noise sensitivity. This evaluation relies on the fact that in four of the five publications which showed correlation coefficients, the fear variable correlated more strongly with the individual annoyance than did the acoustic variable. Similarly, Fields [87] considered the fear variable to be the most important non-acoustic variable related to annoyance in most of the 282 field studies on noises of different origin. It should be noted, however, that the theoretical status of the fear variable is not always clear. For instance, Leonard & Borsky [80] present a path model (i.e., the result of a statistical causal analysis of empirical data) using fear as mediating annoyance, while in several other studies, the data fit better to fear as a moderator.

4.3.3 Evaluation of the source

A related but distinct topic is the general attitude to the source. This is partly a social factor, because large groups of noise-exposed or non-exposed residents share the common attitude that some sources are more tolerable than others. For instance, most people believe railroad traffic to be somehow "better" than road traffic or aircraft transportation. However, I also like to discuss this potentially moderating variable under the heading of "personal factors", because there is sometimes considerable variance of evaluations between residents affected by the same source.

We know both from everyday experience as well as controlled experiments, that persons who are convinced of the importance and necessity of the source show less noise annoyance than persons who are not convinced of the importance. In one of the earliest military aircraft noise studies, Borsky [77] found correlations between source evaluation and noise annoyance in the order of –0.25; this covariation is still higher with annoyance due to private airplanes [89].

4.3.4 Capacity to cope with noise

The theory of psychological stress, as proposed by Lazarus [90, 91] states that psychological stress is the consequence of a person's inability to effectively cope with demands from the environment. Central to the coping concept is the belief and confidence of an affected person that he/she will somehow contend with the problem. This coping strategy can be direct (e.g., in turning off the noise source, or negotiating with the people responsible for the stress) or indirect (mostly via cognitive control, e.g., by means of an exact knowledge of the time schedule of the noise source). Generally, environmental noise sources cannot be turned off directly, but they can be negotiated, and indirect coping strategies can also be very effective in reducing the noise annoyance. Most readers will be familiar with the laboratory experiments of Glass & Singer [92] showing significantly less noise effects with subjects who either knew the time structure of the noise, or knew that they could eventually stop the noise.

In an attempt to measure the self-rated coping capacity of subjects exposed to environmental noise, Guski *et al.* [84] constructed a scale, containing items that reflect successful or unsuccessful coping strategies, such as "If it is too loud outside, I simply close the windows, and then I am no longer disturbed", or "I know that I can protect myself quite well against noise". The sum of six items turned out to be the most efficient moderator in a field study on the effects of seven different noise sources [31].

5 Community actions against noise

According to Fidell [5], an alternative definition of community response to noise is "what the community does about noise or sources", and he mentions local noise ordinance as the main tool of community policy against noise. One of the lines followed by local authorities is to restrict individual noise sources with respect to number, level, and time of day and this line is mainly used against individual noise sources, such as workshops, or construction sites. Very rarely, the same policy is used against regional or national noise sources, such as road, railroad or aircraft traffic. Instead, the main policy against such public sources is (a) to construct noise barriers, (b) to install sound insulating windows, and (c) to separate dwelling units from noise sources.

The main purpose of such actions against noise should be to reduce noise effects, e.g., to reduce the annoyance and disturbance, but there are extremely few studies evaluating the noise abatement programs. Mostly, authorities are happy with showing the sound energy decreasing by some decibels. From the perspective of residents living behind a noise barrier, or a sound insulating window, things may be more complicated: residents may be happy that there has been "something done at least", they may complain because of the ugly rear view of the noise barrier, or because they cannot communicate with their children playing outside the closed window, or because they feel it difficult to control the air exchange in a sound insulated room. Altogether, they may be less annoyed by the new living conditions (noise and noise abatement) than before - or they may not.

One of the earliest studies of the effects of noise barriers on residents [94] has already mentioned the possible conflict between noise reduction and aesthetic disfigurement: the (rather plain) experimental noise barriers did in fact reduce the L_{A10} about 8 dB at the first floor of the houses, and two-thirds of the residents reported the barrier was a good idea, mainly because it reduced noise. About 10% of the respondents complained that it cut off the view of the traffic, and that it was unsightly. About 70% of the people living 20 m directly behind the barrier proposed improving the aesthetic appearance of the barrier by means of a hedge.

An early review of different noise abatement procedures by Kastka [95] showed that different methods to reduce the noise may have different effects on residents, even if the noise level is reduced by comparable levels. For example, a program for traffic calming in a residential area usually has a smaller acoustic energy reduction effect than a noise barrier, but residents in a comparison study felt much greater relief and less noise annoyance after implementation of the traffic calming than after implementation of the noise barrier. A similar conclusion was drawn by Penn-Bressel [96] in comparing results from different road traffic noise abatement programs. According to her analysis, reducing the L_{Aeq} by 1 dB could either lead to 0.5% of less highly annoyed residents, or to 2.3% less highly annoyed residents. In the latter case, the noise reduction program involved a reduction of the number of heavy vehicles on a residential street; the former case did not involve any reduction of heavy vehicles. The author also pointed to the fact that noise barriers do decrease loudness judgments proportionate to the level decrease, but they do not decrease annoyance judgments proportionately: a level decrease of 6 dB usually leads to annoyance judgments comparable with 2 dB less steady-state level.

In 1995, Kastka *et al.* [97] published one of the few long-term studies on the effects of noise barriers near motorways on residents. They compared two acoustic and psychological surveys at four different noise barriers within four adjacent research areas and twelve experimental sites, and one untreated control area with two sites, in the outskirts of the cities Düsseldorf, Wuppertal, Krefeld and Neuss in 1976 and 1988. The following conclusions were drawn: (1) There is no simple causal relation between noise level reduction and annoyance reduction. (2) Barriers produce high annoyance reduction at nearby sites but only minimal effects beyond 150 m from the highway. (3) The long-term annoyance reduction, on average, is relatively greater than the noise level reduction would lead us to expect from steady-state studies. (4) After barrier construction the influence of noise level on annoyance is weaker than before construction.

The aesthetic impression of noise barriers has already been mentioned as a potential moderator of the resident's satisfaction. This aspect has been assumed to work in a study comparing noise annoyance in a German and a Swiss town of the same traffic noise energy level [98]: Since the Swiss streets were visually much more attractive than the German streets (containing balconies with flowers etc.), and the annoyance response of Swiss residents were much lower than that of German residents, it was hypothesized that the aesthetic appearance of a street increases the general well-being of residents, and this in turn reduces the annoyance response to road traffic noise. Aesthetic effects are also believed to work following Japanese experiments comparing annoyance judgments of residents: For example, Tamura [99] asked laboratory subjects to judge the noise annoyance of traffic noise at different levels, and he combined the auditory presentations with photos showing streets with varying amounts of vegetation. It turned out that noise annoyance judgments were effectively moderated by the visual impression of the plants. Similar laboratory effects are reported by Viollon & Lavandier [100].

Tamura [99] went a step further and asked foot passengers on a city street to scale several aspects of the environment twice: once in summer, and another time in winter. The street had trees, which lost their green during winter. It turned out that winter judgments on six of eight scales of the Semantic Differential were significantly more negative than summer judgments. He also showed that noise annoyance on streets and places containing vegetation is comparable to noise annoyance on streets and places without vegetation and 5 dB less daytime traffic noise levels.

On the other hand, a recent study casts some doubt on the simple assumption that the mere presence of vegetation may effectively reduce noise annoyance: Watts *et al.* [101] asked volunteer listeners who were taken to different sites near main roads to scale "noisiness" after listening to the road traffic noise for 20 seconds. The roadsites were covered visually by vegetation (trees, bushes, shrubs etc.) between 30 and 90%. The sound pressure level, measured at the listeners, varied between 55 and 70 dB L_{Aeq}. It turned out that (a) the "noisiness" ratings had a very significant linear correlation with sound levels, and (b) the greater coverage of the roadsites did increase (!) the noisiness judgments. Although it is questionable whether the results of such experiments can be transferred to residents in noisy living areas, because listeners in such experimental tasks probably can do nothing else than judge "loudness" - even if we ask them to judge "annoyance" or "noisiness" - it would be naive to believe that mere belts of vegetation can effectively reduce negative noise effects on residents.

On the other hand, the main bulk of data gathered from "real" residents point to the probability that the aesthetic design of residential areas and noise barriers can positively moderate noise effects or help to increase the benefit from noise abatement programs.

References

[1]   Lambert, J. & Vallet, M., *Study related to the preparation of a communication on a future EC noise policy.* Bruxelles: LEN Report No.9420, 1994.

[2]   Umweltbundesamt & Statistisches Bundesamt (eds.), *Umweltdaten Deutschland.* Berlin: Umweltbundesamt, 1998.

[3]   European Commission, *Future Noise Policy. Green paper of the European Committee.* Bruxelles, 1996.

[4]   Ullrich, S., Die Lärmbelastung durch den Straßenverkehr - Folgerungen aus der Bundesverkehrszählung 1995. *Zeitschrift für Lärmbekämpfung,* **45**, pp. 22-26, 1997.

[5]   Fidell, S., Community response to noise. In: D.M. Jones & A.J. Chapman (eds.): *Noise and Society,* pp. 247-277. Chichester: Wiley, 1984.

[6]   Eiff, A.W., Neus, H. & Otten, H., *Prospektive epidemiologische Feldstudie zu Verkehrslärm und Hypertonie-Risiko.* Bonn: Medizinische Universitätsklinik, im Auftrag d. Umweltbundesamtes, 1985.

[7]   Babisch, W., Elwood, P. & Ising, H., Road traffic noise and heart disease risk: Results of the epidemiological studies in Caerphilly, Speedwell and Berlin. *Nizza '93,* pp. 260-267, 1993.

[8]   Eiff, A.W., Czernik, A., Horbach, L., Jörgens, H. & Wenig, H.G., Der medizinische Untersuchungsteil. In: *DFG-Forschungsbericht "Fluglärmwirkungen",* Boppard: Boldt, pp. 349-424, 1974.

[9]   Knipschild, P.V., Medical effects of aircraft noise: Community vascular survey. *International Archives of Occupational and Environmental Health,* **40**, pp. 185-190, 1977.

[10]   Knipschild, P.V. & Oudshoorn, N., Medical effects of aircraft noise: A drug study. *International Archives of Occupational and Environmental Health,* **40**, pp. 197-20, 1977.

[11]   Cohen, S., Evans, G.W., Krantz, D.S. & Stokols, D., Physiological, motivational and cognitive effects of aircraft noise in children. *American Psychologist,* **35**, pp. 231-243, 1980.

[12]   Cohen, S., Krantz, D.S., Evans, G.W. & Stokols, D., Cardiovascular and behavioral effects of community noise. *American Scientist,* **69**, pp. 528-535, 1981.

[13]   Morrell, S., Taylor, R. & Lyle, D., A review of health effects of aircraft noise. *Australian New Zealand Journal of Public Health,* **21**, pp. 221-236, 1997.

[14]   Maschke, C., Harder, J., Hecht, K. & Balzer, H.U., Nocturnal aircraft noise and adaptation. *Noise Effects '98, Congress Proceedings* Vol. 2, pp. 433-438, 1998.

[15]   Evans, G.W., Bullinger, M. & Hygge, S., Chronic noise exposure and physiological response: A prospective study of children living under environmental stress. *Psychological Science,* **9**, pp. 75-77, 1998.

[16]   Ising, H., Rebentisch, E., Poustka, F. & Curio, I., Annoyance and health risk caused by military low-altitude flight noise. *International Archives of*

Occupational and Environmental Health, **62**, pp. 357-363, 1990.

[17] Poustka, F. & Schmeck, K., Über die psychischen Auswirkungen von militärischer Tiefflugtätigkeit auf Kinder. [Psychological effects of military low altitude flight practice on children]. *Zeitschrift für Kinder- und Jugendpsychiatrie,* **18**, pp. 61-70, 1990.

[18] Guski, R., Schuemer, R. & Felscher-Suhr, U., The concept of noise annoyance: how international experts see it. *Journal of Sound and Vibration,* **223**, pp. 513-527, 1999.

[19] Lindvall, T. & Radford, E.P., Measurement of annoyance due to exposure to environmental factors. *Environmental Research* **6**, pp. 1-36, 1973.

[20] Langdon, F.J., Noise nuisance caused by road traffic in residential areas: Part I. *Journal of Sound and Vibration,* **47**, pp. 243-263, 1976.

[21] Langdon, F.J., Noise nuisance caused by road traffic in residential areas: Part II. *Journal of Sound and Vibration,* **47**, pp. 265-282, 1976.

[22] Langdon, F.J., Noise nuisance caused by road traffic in residential areas: Part III. *Journal of Sound and Vibration,* **49**, pp. 241-256, 1976.

[23] Tracor Inc. *Community reaction to airport noise. Final report.* Austin (TEX): Tracor Doc No. T-70-AU-7454-U, 1970.

[24] Björkman, M., Åhrlin, U. & Rylander, R., Aircraft noise annoyance and average versus maximum noise levels. *Archives of Environmental Health,* **47**, pp. 326-329, 1992.

[25] Fields, J.J., DeJong, R.G., Flindell, I.H., Gjestland, T., Job, R.F.S., Kurra, S., Schuemer-Kohrs, A., Vallet, M. & Yano, T., Recommendation for shared annoyance questions in noise annoyance surveys. *Noise Effects '98,* Vol. 2, pp. 481-486, 1998.

[26] Felscher-Suhr, U., Guski, R. & Schuemer, R., Some results of an international scaling study and their implications on noise research. *Noise-Effects '98. 7th International Congress on Noise as a Public Health Problem,* Vol. 2, pp. 733-736, 1998.

[27] Schultz, T.J., Synthesis of social surveys on noise annoyance. *Journal of the Acoustical Society of America,* **64**, pp. 377-405, 1978.

[28] Oliva, C., *Kurzbericht über die akustische und soziologische Feldstudie. Populärfassung der Lärmstudie 90: Belastung und Betroffenheit der Wohnbevölkerung durch Flug- und Strassenlärm in der Umgebung der internationalen Flughäfen der Schweiz.* Schlieren-Zürich: Büro Dr. Carl Oliva, 1993.

[29] Gjestland, T., Liasjö, K., Granöien, I. & Fields, J.M., *Response to noise around Oslo Airport Fornebu.* Trondheim: Elab-Runit Sintef Gruppen. Acoustics Research Center. Report STF40 A90189, 1990.

[30] Fields, J.M. & Walker, J.G., The response to railway noise in residential areas in Great Britain. *Journal of Sound and Vibration,* **85**, pp. 177-255, 1982.

[31] Finke, H.O., Guski, R. & Rohrmann, B., *Betroffenheit einer Stadt durch Lärm. Bericht über eine interdisziplinäre Untersuchung.* Berlin: Umweltbundesamt / Braunschweig: PTB, 1980.

[32] Lazarus, H. A model of speech communication and its evaluation under

disturbing conditions. In: A. Schick, H. Höge & G. Lazarus-Mainka (eds.), *Contributions to psychological acoustics. Results of the 4th Oldenburg Symposium*, pp. 155-184. Oldenburg: Bibliotheks- & Informationssystem der Universität, 1986.

[33]   Olivier-Martin, N., Schieber, J.P. & Muzet, A., Reponses a un questionnaire sur la forme diurne au cours d'une experience de perturbations du sommeil par 4 types de bruits d'avions. *Bulletin de Psychologie,* **26**, pp. 972-994, 1973.

[34]   Finke, H.O. Messung und Beurteilung der "Ruhigkeit" bei Geräuschimmissionen. *Acustica,* **46**, pp. 141-148, 1980.

[35]   Scharnberg, T. Sleep impairments caused by road traffic noise in cities. *Proceedings of Inter-Noise '85*, pp. 953-956, 1985.

[36]   Muzet, A., Schieber, J.P., Olivier-Martin, N., Ehrhart, J. & Metz, B., Relationship between subjective and physiological assessments of noise-disturbed sleep. *Proceedings of the International Congress on Noise as a Public Health Problem*, Dubrovnik, pp. 575-586, 1973.

[37]   Schönpflug, W. & Schulz, P., *Lärmwirkungen bei Tätigkeiten mit komplexer Informationsverarbeitung. Feldstudien in einem Industriebetrieb und Laboruntersuchungen.* Berlin: Freie Universität / Umweltbundesamt., 1979.

[38]   Battmann, W. & Schönpflug, W., Individual noise reduction. In: A. Schick, H. Höge & G. Lararus-Mainka (eds.), *Contributions to psychological acoustics. Results of the 4th Oldenburg Symposium*, pp. 223-237. Oldenburg: Bibliotheks- & Informationssystem der Universität, 1986.

[39]   Ando, Y., Effects of daily noise on fetuses and cerebral hemisphere specialization in children. *Journal of Sound and Vibration,* **127**, pp. 411-417, 1988.

[40]   Cohen, S., Krantz, D.S., Evans, G.W. & Stokols, D., Community noise, behavior, and health: The Los Angeles Noise Project. In: A. Baum & J.S. Singer (eds.), *Advances in Environmental Psychology, Vol. 4: Environment and Health*, pp. 295-317. Hillsdale, NJ: Erlbaum, 1982.

[41]   Hygge, S., Evans, G.W. & Bullinger, M., The Munich airport noise study: Cognitive effects on children from before to after the change over of airports. *Inter-Noise '96*, Vol. 5, pp. 2189-2194, 1996.

[42]   Meis, M., Hygge, S., Evans, G.W., Bullinger, M. & Schick, A., Effects of traffic noise on implicit and explicit memory: Results from field and laboratory studies. *Noise Effects '98, 7th International Congress on Noise as a Public Health Problem*, Vol. 1, pp. 389-394, 1998.

[43]   Haines, M.M., Stansfeld, S.A., Job, R.F.S. & Berglund, B., Chronic aircraft noise exposure and child cognitive performance and stress. *Noise-Effects '98: 7th International Congress on Noise as a Public Health Problem*, Vol. 1, pp. 329-335, 1998.

[44]   Regecová, V. & Kellerová, E., Effects of urban noise pollution on blood pressure and heart rate in preschool children. *Journal of Hypertension,* **13**, pp. 405-412, 1995.

[45]   Müller, F., Pfeiffer, E., Jilg, M., Paulsen, R. & Ranft, U., Effects of acute and chronic traffic noise on attention and concentration of primary school children. *Noise-Effects '98, 7th International Congress on Noise as a Public Health Problem*, Vol. 1, pp. 365-368, 1998.

[46]   Lercher, P., Schmitzberger, R. & Kofler, W., Perceived traffic air pollution, associated behavior and health in an alpine area. *Science of the Total Environment*, **169**, pp. 71-74, 1995.

[47]   Wicke, L., *An environmental balance sheet for the FRG*. Helsinki Symposium on the Benefits of Environmental Policies. Paris: OECD-Organization for Economic Cooperation and Development, 1987.

[48]   Weinberger, M., Kosten des Lärms. *UmweltWirtschaftsForum*, **3**, pp. 38-41, 1995.

[49]   Vallet, M., Annoyance after changes in airport noise environment. *Inter-Noise '96*, Vol. 5, pp. 2329-2334, 1996.

[50]   Kastka, J., Borsch-Galetke, E., Guski, R., Krauth, J., Paulsen, R., Schuemer, R. & Oliva, C., Longitudinal study on aircraft noise. Effects at Düsseldorf Airport 1981-1993. *Proceedings of the 15th International Congress on Acoustics (ICA95)*, Trondheim, Vol. IV, pp. 447-451, 1995.

[51]   Brown, A.L., Hall, A. & Kyle-Little, J., Response to a reduction in traffic noise exposure. *Journal of Sound and Vibration*, **98**, pp. 235-246, 1985.

[52]   Öhrström, E., Community reactions to railway traffic - effects of countermeasures against noise and vibration. *Inter-Noise '97*, Vol. 2, pp. 1065-1070, 1997.

[53]   Job, R.F.S., Over-reaction to changes in noise exposure: The possible effect of attitude. Letter to the editor. *Journal of Sound and Vibration*, **126**, pp. 550-552, 1988.

[54]   Raw, G.J. & Griffiths, I.D., The effect of changes in aircraft noise exposure. Letter to the editor. *Journal of Sound and Vibration*, **101**, pp. 273-275, 1985.

[55]   Griffiths, I.D. & Raw, G.J., Adaptation to changes in traffic noise exposure. *Journal of Sound and Vibration*, **132**, pp. 332-336, 1989.

[56]   Hatfield, J. & Job, R.F.S., Evidence of optimism bias regarding the health effects of exposure to noise. *Noise-Effects '98, 7th International Congress on Noise as a Public Health Problem*, Vol. 1, pp. 251-254, 1998.

[57]   Oliva, C., *Belastung der Bevölkerung durch Flug- und Straßenlärm. Eine Lärmstudie am Beispiel der Flughäfen Genf und Zürich*. Berlin: Duncker & Humblot, 1998.

[58]   Botteldoren, D.B., Decloedt, S., Bruyneel, J. & Pottie, S., Characterisation of quiet areas: subjective evaluation and sound level indices. 137th ASA Meeting / Forum Acusticum 1999 / 25. DAGA Conference. Abstract in *Acta Acustica*, **85**, Suppl. 1, S451, 1999.

[59]   Fields, J.M. & Walker, J.G., Comparing the relationships between noise level and annoyance in different surveys: A railway noise vs. aircraft and road traffic comparison. *Journal of Sound and Vibration*, **81**, pp. 51-80, 1982.

[60]   Schuemer, R. & Schuemer-Kohrs, A., Lästigkeit von Schienenverkehrslärm im Vergleich zu anderen Lärmquellen - Überblick über Forschungsergebnisse. *Zeitschrift für Lärmbekämpfung,* **38**, pp. 1-9, 1991.

[61]   Schreckenberg, D., Schuemer, R., Schuemer-Kohrs, A., Griefahn, B. & Moehler, U., Attitudes toward noise source as determinants of annoyance. *Euro-Noise 98,* Vol. 1, pp. 595-599, 1998.

[62]   Guski, R., Psychological determinants of train noise annoyance. *Euro-Noise 98,* Vol. 1, pp. 573-576, 1998.

[63]   Job, R.F.S. Community response to noise: A review of factors influencing the relationship between noise exposure and reaction. *Journal of the Acoustical Society of America,* **83**, pp. 991-1001, 1988.

[64]   Taylor, S.M., A comparison of models to predict annoyance reactions to noise from mixed sources. *Journal of Sound and Vibration,* **81**, pp. 123-138, 1982.

[65]   Diamond, I.D. & Rice, C.G., Models of community reaction to noise from more than one source. In: H.S. Koelega (ed.), *Environmental annoyance: Characterization, measurement, and control. Proceedings of the International Symposium on Environmental Annoyance,* Woudschoten (NL), pp. 301-310. Amsterdam: Elsevier, 1987.

[66]   Izumi, K., Annoyance due to mixed source noises. A laboratory study and field survey on the annoyance of road traffic and railroad noise. *Journal of Sound and Vibration,* **127**, pp. 485-489, 1988.

[67]   Miedema, H.M.E. & Van den Berg, R., Community response to tramway noise. *Journal of Sound and Vibration,* **120**, pp. 341-346, 1988.

[68]   Ronnebaum, T., Schulte-Fortkamp, B. & Weber, R., Evaluation of combined noise sources. In: A. Schick & M. Klatte (eds.), *Contributions to Psychological Acoustics. Results of the 7th Oldenburg Symposium on Psychological Acoustics.* Oldenburg: BIS, pp. 171-189, 1997.

[69]   Guski, R. Interference of activities and annoyance by noise from different sources: Some new lessons from old data. In: A. Schick & M. Klatte (eds.), *Contributions to Psychological Acoustics. Results of the 7th Oldenburg Symposium on Psychological Acoustics,* pp. 239-258. Oldenburg: BIS, 1997.

[70]   Schulte-Fortkamp, B., The inherent context of annoyance ratings on combined noises. *Noise Effects '98, 7th International Congress on Noise as a Public Health Problem,* Vol. 2, pp. 683-686, 1998.

[71]   Kaku, J., Kato, T., Kuwano, S. & Namba, S., Predicting overall reaction to multiple noise sources. 137th ASA Meeting / Forum Acusticum / 25th DAGA Conference. Abstract in *Acta Acustica,* **85**, Suppl. 1, S79, 1999.

[72]   Baron, R.M. & Kenny, D.A., The moderator-mediator variable distinction in social psychological research: conceptual, strategic, and statistical considerations. *Journal of Personality and Social Psychology,* **51**, pp. 1173-1182, 1986.

[73]   Fields, J.M. *The timing of noise-sensitive activities in residential areas.* Hampton (Virginia): Nasa Contractor Report 177937, 1985.

[74]   Guski, R., Probst, W., Neuschwinger, B., Schlebusch, P., Van den Brulle,

P. & Gerlinger, H., *Störwirkungen von Sportgeräuschen im Vergleich zu Störwirkungen von Gewerbe- und Arbeitsgeräuschen. Eine interdisziplinäre Felduntersuchung über Freizeit- und Gewerbegeräusche in Wohngebieten.* Bochum und München: im Auftrag des Umweltbundesamtes. Forschungsbericht 105 01317/02, 1989.

[75] Maschke, C., Harder, J., Hecht, K. & Balzer, H.U., Nocturnal aircraft noise and adaptation. *Noise Effects '98, 7th International Congress on Noise as a Public Health Problem,* **2,** pp. 433-438, 1998.

[76] Umweltbundesamt, *Daten zur Umwelt,* Berlin: E. Schmidt Verlag, 1997.

[77] Borsky, P.N. *Community reactions to Air Force noise I: Basic concepts and preliminary methodology. II: Data on community studies and their interpretation.* Chicago: National Opinion Research Center of the University of Chicago. WADD Technical Report 60-689, 1961.

[78] McKennell, A.C., *Aircraft noise annoyance around Heathrow airport.* London: Her Majesty's Stationery Office, 1963.

[79] Schuemer, R., *Fluglärmwirkungen III: Ergänzende Analysen zum sozialwissenschaftlichen Untersuchungsteil des Fluglärmprojektes der DFG.* Boppard: Harald Boldt Verlag, 1974.

[80] Leonard, S. & Borsky, P.N., A causal model for relating noise exposure, psychosocial variables and aircraft noise annoyance. *Proceedings of the International Congress on Noise as a Public Health Problem,* Dubrovnik, pp. 691-705, 1973.

[81] Fidell, S. Why is annoyance so hard to understand? In: H.S. Koelega (ed.), *Environmental annoyance: Characterization, measurement, and control. Proceedings of the International Symposium on Environmental Annoyance,* Woudschoten (NL). Amsterdam: Elsevier, pp. 51-54, 1987.

[82] Schick, A. *Systematische Analyse von schriftlichen Anfragen bei Bürgerberatungsstellen. Schlußbericht.* Oldenburg: Institut zur Erforschung von Mensch-Umwelt-Beziehungen, 1986.

[83] Schick, A. Personality factors in sound perception: Historical developments. *Inter-Noise 97, Proceedings,* **2,** pp. 1027-1032, 1997.

[84] Guski, R., Wichmann, U., Rohrmann, B. & Finke, H.O., Konstruktion und Anwendung eines Fragebogens zur sozialwissenschaftlichen Untersuchung der Auswirkungen von Umweltlärm. *Zeitschrift für Sozialpsychologie,* **9,** pp. 50-65, 1978.

[85] Weinstein, N.D. Individual differences in reactions to noise: A longitudinal study in a college dormitory. *Journal of Applied Psychology,* **63,** pp. 458-466, 1978.

[86] Zimmer, K. & Ellermeier, W., Ein Kurzfragebogen zur ökonomischen Erfassung der Lärmempfindlichkeit. *Umweltpsychologie,* **2,** pp. 54-63, 1998.

[87] Fields, J.M., *Effect of personal and situational variables on noise annoyance: With special reference to implications for en route noise.* Washington, DC: U.S. Department of Transportation. Federal Aviation Administration. Report DOT/FAA/EE-92/03, 1992.

[88] McKennell, A.C. Psycho-social factors in aircraft noise annoyance.

Proceedings of the International Congress on Noise as a Public Health Problem, Dubrovnik, pp. 627-644, 1973.

[89]   Rohrmann, B. Die Störwirkung des Flugbetriebs an Landeplätzen - eine empirische Studie. *Kampf dem Lärm,* **23**, pp. 6-11, 1976.

[90]   Lazarus, R.S. & Launier, R., Stress-related transactions between person and environment. In: L. Pervin & M. Lewis (eds.), *Perspectives in Interactional Psychology,* New York: Plenum, pp.1-67, 1978.

[91]   Lazarus, R.S. *Emotion and Adaptation.* New York: Oxford University Press, 1991.

[92]   Glass, D.C. & Singer, J.E., *Urban Stress. Experiments on Noise and Social Stressors.* New York: Academic Press, 1972.

[93]   Flindell, I.H. Military aircraft noise and health -- Methodological issues in research, 137th ASA Meeting / Forum Acusticum / 25th DAGA Conference. Abstract in *Acta Acustica,* **85**, Suppl. 1, pp. S351, 1999.

[94]   Scholes, W.E., Mackie, A.M., Vulkan, G.H. & Harland, D.G., Performance of a motorway noise barrier at Heston. *Applied Acoustics,* **7**, pp. 1-13, 1974.

[95]   Kastka, J. Untersuchungen zur subjektiven Wirksamkeit von Maßnahmen gegen Verkehrslärm und deren Moderatoren durch nichtakustische Faktoren, In: H. Haase & W. Molt (eds.), *Handbuch der Angewandten Psychologie.* Verlag Moderne Industrie, **3**, Markt und Umwelt, pp. 468-485, 1981.

[96]   Penn-Bressel, G. Subjektive Wirksamkeit von Lärmschutzmaßnahmen, *Fortschritte der Akustik, DAGA '88,* pp. 213-216, 1988.

[97]   Kastka, J., Buchta, E., Ritterstaedt, U., Paulsen, R. & Mau, U., The long term effect of noise protection barriers on the annoyance response of residents, Journal of Sound and Vibration, **184**, pp. 823-852, 1995.

[98]   Kastka, J., Noack, R.H., Mau, U., Maas, P., Conrad, U., Ritterstedt, U. & Hangartner, M., Comparison of traffic-noise annoyance in a German and a Swiss town: effects of the cultural and visual aesthetic context, In: A. Schick, H. Höge & G. Lazarus-Mainka (eds.), *Contributions to psychological acoustics. Results of the 4th Oldenburg Symposium.* Oldenburg: BIS , pp. 312-340, 1986.

[99]   Tamura, A. Effects of landscaping on the feeling of annoyance of a space., In: A. Schick & M. Klatte (Eds.): *Contributions to Psychological Acoustics. Results of the 7th Oldenburg Symposium on Psychological Acoustics.* Oldenburg: BIS, pp. 135-161, 1997.

[100] Viollon, S. & Lavandier, C., Influence of vision on audition in an urban context, *Inter-Noise 97,* **2**, pp. 1167-1170, 1997.

[101] Watts, G., Chinn, L. & Godfrey, N., The effects of vegetation on the perception of traffic noise, *Applied Acoustics,* **56**, pp. 39-56, 1999.

[102] Lambert, J., Kail, J.M. & Quinet, E., Transportation noise annoyance: An economic issue. *Noise-Effects '98: 7th International Congress on Noise as a Public Health Problem,* Vol.2, pp. 749-754, 1998.

Chapter 5

Prediction of urban noise

Miguel Arana

Physics Department, Public University of Navarra, Pamplona, Spain

1 Introduction

The prediction of noise is a powerful tool for several reasons. As said in medical terms prevention is better than cure, and prior to urban planning an accurate prediction of noise can detect problems to which the subsequent solutions may be either non-viable or very expensive. Accuracy of prediction requires knowledge of sources' characteristics as well as the geometry and conditions of the propagation paths. Levels at receiver points can be compared to either legal limits or quality objectives. Once the prediction model is validated calculating new levels on receiver points, due to variations in sources or propagation paths, is easy and fast. Furthermore the powerful graphic programs used for showing results allow a clear display of them over the whole selected area.

This chapter is organized in the following manner. In section 2, advantages and disadvantages between traditional measurements and computational models based on standards are commented on. Basic acoustic definitions about noise sources (sound power and directivity pattern) are introduced in section 3. Section 4 deals with factors affecting outdoor sound propagation. Section 5 reports some characteristics of the most important sources of urban noise. Section 6 deals with the prediction of urban noise including a basic theory for the calculation of traffic noise, standards as well as a brief comparison between them. Section 7 deals with vehicle noise regulations and their evolution. The prognosis for urban noise evolution is briefly discussed in section 8.

2 What is urban noise prediction for?

Traditionally, acoustic measurements had been carried out as the technique in order both to quantify and to solve environmental noise problems. At present simple equipment to measure many acoustic parameters and indices are available. Measurements with such equipment are fast and easy. There is no doubt that measurements will always be necessary. However, measurements are very dependent on meteorological conditions. If it is raining you can not measure. If it is very windy either you can not measure or you are not sure if the measured noise corresponds to the sources of interest. Results are sensitive to background noise and may require other sources to be shut off. Finally, limitations in measurement points can occur.

Computational models for the prediction of urban noise caused by different sources (industry, road traffic, aircraft) can offer some advantages. They give a greater level of detail both in terms of frequency, source and receiver point. They are well suited as a planning tool and they can be easily updated. Both changes and additional data about sources of noise allow having the models updated. Against this, an extensive data collection (noise and geometry) is required to use them as well as higher acoustical skills.

The basic formula for the prediction of noise is quite simple. Basically the predicted level comes from the characteristics of the source (acoustic power and directivity) minus the attenuation between the source and the reception point:

$$L_{Pr} = L_W + DF + A \qquad (1)$$

where:

L_{Pr} : sound pressure level at the reception point
L_W : acoustic power of the source
DF : directivity factor in the direction source-receiver
A : correction factors (spreading, absorption, reflections, screenings, etc.)

If there are many sources one calculation for each one is required. The resultant level can be obtained by addition of the partial levels. However, that simple equation requires a detailed knowledge both of acoustic properties of emission of the source and of geometry of the propagation paths. For stationary sources (typically industrial noise) measurements to characterize the source can be achieved in a simple arrangement. For non-stationary sources (e.g. road traffic) it is not possible in practice to characterize each and all of the individual sources. In these cases, standards provide either the sound emission of one vehicle or the sound emission from a road in dBA/m, normally as a function of its speed for each vehicle class.

3 Sound power and directivity pattern of noise sources

The first step in order to predict noise is to know the characteristics of the sources of noise. The two necessary attributes to describe completely the strength of a noise source are its sound power level and its directivity pattern, both usually stated as a function of frequency. The sound power quantifies the total sound power W, in watts, radiated by the source in all directions. That is a typical characteristic of each source, regardless of the environment in which it operates. It is usually stated in octave bands. Using $W_0 = 10^{-12}$ W as the reference sound power, the sound power emission level is defined as:

$$L_W = 10 \log \frac{W}{W_0} \qquad (2)$$

When A-weighting is applied to the overall sound power emission level the result is expressed as L_{WA}. For instance, for a source radiating 2 W, the sound power emission level is $L_W = 123$ dB. The sound pressure level, L_P, is

$$L_P = 10 \log \frac{P^2}{P_0^2} \qquad (3)$$

where P^2 is the mean-square sound pressure and P_0^2 is the reference mean-square sound pressure, being P_0 standardized at 20 µPa for airborne sound. If the source radiates omnidirectionally with sound power W, the sound pressure level at distance r, $L_P(r)$, is

$$L_P(r) = 10 \log \frac{W \cdot 10^{12}}{4\pi r^2} \qquad (4)$$

because of $I = P^2/\rho c = W/4\pi r^2$ and assuming $\rho c = 400$ mks rayls, ρ being the density of air and c the speed of sound in air. The majority of sources are directional. It means that different levels are found at the same distance for different directions. If attention is restricted to the far field [1] the radiation from any source can be expressed as a product of an on-axis pressure $P_{ax}(r)$ which depends only on r and a term $H(\theta, \phi)$, which depends only on direction defined by angles θ and ϕ. $H(\theta, \phi)$ is called the *directional factor* and it is always normalized so that its maximum value is unity. Such direction determines the *acoustic axis*. The radiated sound power can be obtained by integration of the intensity I versus pressure over a surface, a sphere of radius r, enclosing the source

$$W = \frac{1}{2\rho c} \int_{4\pi} P^2(r, \theta, \phi) \, r^2 d\Omega = \frac{1}{2\rho c} r^2 P_{ax}^2(r) \int_{4\pi} H^2(\theta, \phi) d\Omega \qquad (5)$$

For an omnidirectional source, the pressure amplitude is not dependent on the direction. Denoting by $P_S(r)$ its pressure amplitude at distance r and assuming the same acoustic power generated, it can be expressed as

$$W = \frac{1}{2\rho c} 4\pi r^2 P_S^2(r) \qquad (6)$$

A numerical measurement of the directivity of a sound source is the *directivity factor*. The directivity factor $Q_{\theta,\phi}$ is defined as the ratio of (1) the mean-square sound pressure at distance r and on the direction defined by angles θ and ϕ from a source radiating W watts to (2) the mean-square sound pressure at the same distance from an omnidirectional source radiating the same acoustic power W. Assuming that the directivity pattern does not change shape regardless of the radius r at which it is measured, we get

$$Q_{\theta,\phi} = \frac{P^2(\theta,\phi)}{P_S^2} = \frac{I(\theta,\phi)}{I_S} \tag{7}$$

$Q_{\theta,\phi}$ is a dimensionless quantity. For a directional source with directivity factor $Q_{\theta,\phi}$ the sound pressure level, $L_P(r,\theta,\phi)$, at distance r in the direction defined by angles θ and ϕ is

$$L_P(r,\theta,\phi) = 10\log\frac{W \cdot Q_{\theta,\phi} \cdot 10^{12}}{4\pi r^2} \tag{8}$$

The *directivity index*, $DI_{\theta,\phi}$, which is the decibel equivalent of the directivity [2], is simply defined as

$$DI_{\theta,\phi} = 10\log Q_{\theta,\phi} \tag{9}$$

Now, eqn (8) can be written as

$$L_P(r,\theta,\phi) = L_W + DI_{\theta,\phi} - 20\log r + 109 \tag{10}$$

taking 20 µPa and 1 pW as the reference sound pressure and reference sound power respectively. As a general rule, the majority of actual sources are omnidirectional for very low frequencies and directional for high frequencies. For sources strongly directional the calculation of the sound pressure level is very different if estimating outdoors or indoors. In closed spaces and for directive sources, the calculation of the sound pressure levels may be sufficiently accurate by considering only the intensity on the acoustic axis and the angle width of the principal lobe. This width is usually defined by a fall of 3 dB with regard to the level on the acoustic axis.

From the early 1980s commercial sound intensity measurement systems were available on the market. Curiously with a 50 years' delay with respect to the first patent for a device for the measurement of sound energy flux - granted to Harry Olson of the RCA Company in America in 1932 [3]. It was made possible with the convergence of theoretical and experimental advances, including the derivation of the cross-spectral formulation for sound intensity measurements from the laboratory into practical use [4]. With regard to urban noise control, two of the most common practical applications of sound intensity measurement are the sound power determination and noise source identification. By this means, different parts of a complex source can be identified as well as quantified in their sound power [5].

4 Outdoor sound propagation

4.1 Introduction

There are three basic components in the outdoor propagation of sound: source, path and receiver. Notion of source can include many sources with many different characteristics both in sound power and directivity. Similarly notion of path can include several paths with differences in both distances traveled and reflection or absorption processes. Finally, the receiver could be just one point or a wide area. At least in a theoretical way the sound level at the receiver can be calculated by combining partial levels travelling through all paths from all significant sources. In this section outdoor sound propagation along the path from source to receiver is briefly discussed.

4.2 Geometric spreading of sound

Taking into account only the geometrical spreading of noise from a non-directional point source, eqn (4) yields the sound pressure level at distance r according to the law of energy conservation. For a directional source with directivity factor $Q_{\theta,\phi}$ eqn (10) yields the sound pressure level at distance r in the direction defined by angles θ and ϕ.

4.3 Ground attenuation

The effect of ground on sound level at the receiver is due to the interaction between the direct source-to-receiver sound and the ground reflected ray [6]. This interference may produce either attenuation or amplification with regard to the direct sound. The amplitude and the phase of the reflected wave at the receiver point depend both on the path length difference and on the acoustic impedance of the ground. It is clear that a maximum amplification of 6 dB respecting the direct sound level can be expected at frequencies for which both rays are in phase and the path difference is small. On the other hand, attenuations between 20-30 dB at some frequencies can occur in audio-frequency range with suitable conditions [7].

Accurate calculations upon ground correction need several inputs both for geometry and for ground characteristics. Standards usually simplify this effect by assuming two types of ground: hard ground (asphalt, water) and soft ground (grassland, ploughed earth, snow). Furthermore, spreading and ground effect are usually jointly handled into the distance adjustment heading.

A typical formula relating the distance adjustment connecting the ground effect with spreading can be written as:

$$\text{Attenuation} = 10 \log \left(\frac{D_0}{D} \right)^{1+\alpha} \tag{11}$$

where D_0 is the distance of reference

  D is the distance from source to receiver

  α is the absorption coefficient, dependent on type of ground.

Fig. 1 shows attenuation by distance and by different values of α, according to eqn (11). In this example $D_0 = 15$ meters.

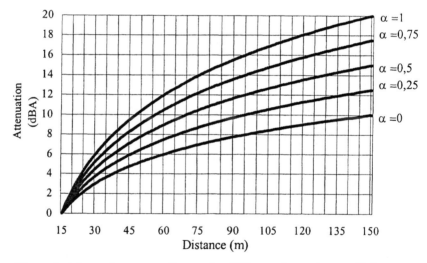

Figure 1: Attenuation versus distance by different absorption coefficients.

4.4 Attenuation of outdoor sound propagation by atmospheric absorption

When a sound wave travels through air all acoustic energy is ultimately degraded into thermal energy. There are three basic mechanisms for the dissipation: viscosity, heat conduction and molecular relaxation. The first successful theory of sound absorption was developed by Stokes [8]. Kirchhoff utilized the property of *thermal conductivity* to develop an additional contribution [9]. However, additional microscopic absorption mechanisms occur both within and between molecules [10].

A simple way to introduce a loss mechanism relating to atmospheric absorption is by means of the *relaxation time* which allows of a delay between pressure and condensation. Assuming an exponential decay by acoustic pressure in function of distance: $P(r) = P_0 \cdot e^{-\alpha r}$, the acoustic intensity at distance r, I(r) can be expressed by:

$$I(r) = \frac{\left(P_0 \cdot e^{-\alpha r}\right)^2}{2\rho c} = I_0 \cdot e^{-2\alpha r} \qquad (12)$$

where I_0 is the acoustic pressure at the initial point and α is the *absorption coefficient*. In eqn (12) c is the phase speed and it is a function of frequency. In such a case the propagation is termed *dispersive*. Absorption coefficients for waves decaying with distance are frequently expressed in *nepers per meter,* neper (Np) being a dimensionless unit. Note that when $r = 1/\alpha$ the pressure amplitude has dropped to 1/e of its initial value P_0. In terms of acoustic pressure, the attenuation A_a in dB, due to atmospheric absorption is given by:

$$A_a = 10 \log \frac{I_0}{I(r)} = 10 \log(e^{2\alpha r}) = 20 \log(e^{\alpha r}) \approx 8.7 r \alpha \qquad (13)$$

Consequently, a = 8.7α is a measure of the spatial rate of decrease in sound pressure level expressed in dB/m.

Table 1. Attenuation coefficient a (dB/km)

Temperature ° C	Relative humidity (%)	Frequency (Hz)							
		62.5	125	250	500	1000	2000	4000	8000
0	10	0.42	1.30	4.00	9.25	14.0	16.6	19.0	26.4
	50	0.18	0.41	0.82	2.08	6.83	23.8	71.0	147
	90	0.13	0.37	0.76	1.45	3.66	12.1	43.2	138
10	10	0.34	0.79	2.29	7.52	21.6	42.3	57.3	69.4
	50	0.16	0.49	1.05	1.90	4.26	13.2	46.7	155
	90	0.10	0.35	1.00	2.00	3.54	8.14	25.7	92.4
20	10	0.37	0.78	1.58	4.25	14.1	45.3	109	175
	50	0.12	0.45	1.32	2.73	4.66	9.86	29.4	104
	90	0.07	0.27	0.97	2.71	5.30	9.06	20.2	62.6
30	10	0.36	0.96	1.82	3.40	8.67	28.5	96.0	260
	50	0.09	0.35	1.25	3.57	7.03	11.7	24.5	73.1
	90	0.05	0.20	0.78	2.71	7.32	13.8	23.5	53.3

Absorption coefficient depends mainly on the frequency of sound but also on temperature, humidity and atmospheric pressure [11]. Values of the absorption coefficient in respect both of temperature and of relative humidity are given in Table 1 [12] for different discrete frequencies.

As can be seen in Table 1, atmospheric absorption is very small for low and medium frequencies. Because the most important source of urban noise is road traffic (with predominant emission in such frequency range) this effect is either neglected or approximated by small corrections in standards.

4.5 Barriers

A barrier (or screen) is any obstacle intercepting the line of sight from source to receiver. A barrier reduces the sound level at the receiver point. It is a diffraction problem [13] and consequently attenuation depends strongly on frequency. First it is assumed that transmission loss through the barrier is higher than the potential screening performance determined by the diffraction effect from the barrier geometry.

Several experiments had been carried out on this subject [14], [15], [16]. A simplified formula determining approximately the attenuation of sound A_b in dB, from a point source by an infinitely long thin barrier is given by:

$$A_b = 10 \log (3 + 20 \ N) \qquad (14)$$

where $N = \pm \dfrac{2}{\lambda}(a + b - c)$ is the Fresnel number (see Fig. 2), λ is the wavelength, and a, b and c are in meters.

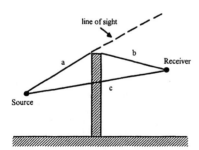

Fig. 2: Ray path geometry for screen diffraction without ground effects.

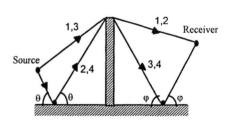

Fig. 3: Ray path geometry for screen diffraction including ground effects.

On the line of sight (N = 0) an attenuation of 5 dB still occurs. Even though A_b increases with N, a practical limit of 20 dB is assumed. Fig. 9 shows potential correction as a function of path difference for the calculation of road traffic noise to UK standards.

Usually source and receiver are located near the ground. Therefore ground reflections must be included in the analysis. As is shown in Fig. 3, four paths are involved and the pressure at the receiver is determined by summing contributions from each diffracted path.

4.6 Vegetation

Bushes and trees (wooded areas in general) can provide attenuation in the propagation of sound. There are two basic effects. At low frequencies and because the roots make the ground more porous, ground attenuation is enhanced [17]. On the other hand, sound scattering both by trunks (middle frequencies) and by leaves (high frequencies) causes attenuation of sound [18]. A noise attenuation with respect to grassland of the order of 0.1 dBA per meter of dense vegetation depth is a realistic estimation for L_{eq}. In respect of frequency, f, the attenuation caused by heavy wood areas can be estimated [19] by:

$$A_w = (6 \cdot \text{to} \cdot 10) \left(\frac{f}{1\,kHz} \right)^{1/3} \left(\frac{r_w}{100\,m} \right) \leq 10 \qquad (15)$$

where r_w is the depth of the wooded area.

Actually, the noise attenuation provided by vegetation is generally overestimated. Probably there is a psychological effect to it.

4.7 Reflection

When there is a substantial reflecting surface on the far side of the traffic flow, reflected sound is added to direct sound and the sound level increases. When there are substantial reflecting surfaces - typically parallel building façades on both sides of the street - urban reverberation is caused because the sound's divergence outward from the roadway is restricted. In the first case, the distance between receiver point and the nearest reflecting surface, d, as well as the absorption coefficient of it, are the parameters influencing sound level increment. In addition to them, the building height, h, and the width of the street, w, are important for urban canyon situations.

Some standards apply a correction of +1 dBA when there is a reflecting opposite façade. For urban canyons reference [19] suggests a formula to calculate the increment on noise level, Ar, for a single lane of traffic due to that effect:

$$A_r = 10 \cdot \log \left[1 + \left(\frac{r}{r+2d} \right)^2 (1 - \alpha) \right] + R \qquad (16)$$

with R = 4 (h/w) $\leq$ 3 dBA; r is the perpendicular distance from traffic to receiver point and the rest of the parameters have been defined above. As an example, for the next values of the parameters: w = 20 m, d = 2 m, r = 8 m, h = 15 m and α = 0.1, an increment of 4.5 dBA is applied by eqn (16).

The effect of a canyon street when noise is emitted by traffic and repeatedly reflected from parallel building façades also depends on the structure and roughness of the façades. More accurate prediction is possible with scale models incorporating realistic scattering elements [20], [21].

5 Sources of urban noise

Noise sources and their characteristics in vehicles, trains and aircraft are briefly described in the next sections.

5.1 Road traffic noise: Characteristics

Each vehicle represents a complex noise source. Actually one vehicle is composed of numerous sources: engine, air inlet and exhaust, transmission, tire/road surface and others. Moreover the power and directivity of these sources depend on other variables such as speed, mode of operation or type of asphalt. In practice the prediction of road traffic noise is not possible only using the characteristics of the emission of the individual vehicles. However, in most situations, we are interested in the noise caused by roads with continuous traffic flow. In these situations the interesting variables are the statistical traffic variables as a whole such as average speed or composition of the traffic. The formulas of prediction allow for the calculation of the basic level at the reference distance in respect of the traffic variables. Computational models start assigning an acoustic power per road unit length. Subsequently, the effects of divergence, absorption, diffraction and reflection for propagating waves are evaluated.

The amount of noise emitted by a vehicle depends mainly on two factors, namely type of vehicle and its speed. In the towns, there are very diverse conditions. A comprehensive data collection on peak noise levels from different categories of vehicles operating in urban traffic conditions is shown in Fig. 4 [22]. The distributions were determined from over 22,000 vehicle pass-by events.

For cars, the power train noise (engine, air inlet, exhaust and transmission) is dominant at low speeds; at higher speeds tire rolling noise and aerodynamic noise is either of the same order or higher than power train noise. The noise is roughly proportional to the sixth power of air speed [23]. For trucks the engine, exhaust and cooling fan noise are the dominant sources under most operating conditions although the noise of the tire rolling on the road surface can be similar at high speeds. Fig. 5 shows a comparison between power train noise and rolling aerodynamic noise at various speeds for cars and trucks both for urban streets and motorways [24].

With regard to noise spectra of cars and trucks moving under freely flowing conditions, Fig. 6 shows a typical shape in octave bands [25]. As is shown in the figure, low and middle frequency ranges predominantly control the A-weighted levels.

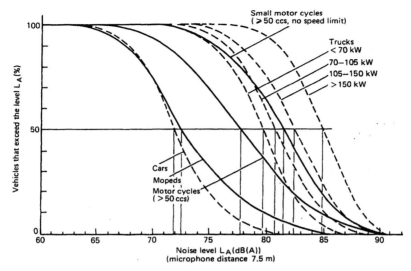

Figure 4: Distribution of peak noise levels of different types of vehicles taken from vehicles operating in urban traffic conditions. (From reference [22].)

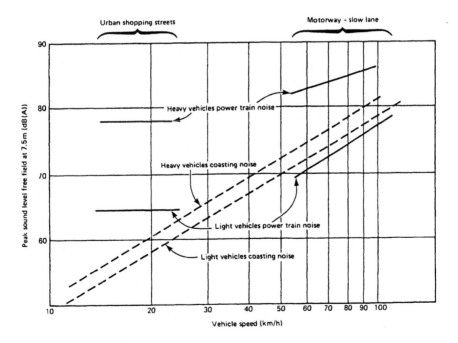

Figure 5: Vehicle power train noise and coasting noise (tire/road surface and aerodynamic noise). (From reference [24].)

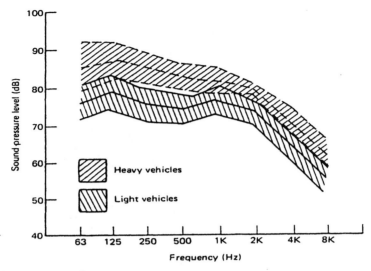

Figure 6: Typical noise spectra of light and heavy vehicles under freely flowing
conditions. (From reference [25].)

5.1.1 Engine noise

The engine noise is originated by the gas loads applied to the engine structure by
the combustion process as well as by the impacts of the pistons against the
cylinder walls, which induce vibrations in the structure. These two sources of
noise are respectively known as *combustion noise* and *mechanical noise*. The
principal sources of engine noise were clearly established in the early 1980s [26],
[27], [28].

For diesel engines the noise level L in dBA increases in proportion about 30
log(N), where N is the engine speed in r.p.m. In the case of spark ignition
engines such a proportion is about 50 log(N) [29]. With regards to the frequency
spectrum the sound level decreases when frequency increases. Typical rates of
decay are 8 dB/octave for diesel engines and 15 dB/octave for petrol engines.

5.1.2 Tire noise

The noise due to tire/road contact is caused by three main reasons: a) vertical
excitation and radiation of noise from the tire casing, b) tangential excitation as a
result of stick and slip action, and c) suction and ejection of air. Tire noise
depends on the tire type and its tread design as well as on the type and texture of
the road surface. Because the noise coming from the exhaust, intake and engine
have been notably reduced, tire noise is now predominant for speeds above 70
km/h in the case of passenger cars. On wet roads, speeds can be much lower and
power spectral density increases with the frequency. Fig. 7 shows the power
spectral density of tire noise measured at 7.5 m by different speeds on an ISO
track surface [30].

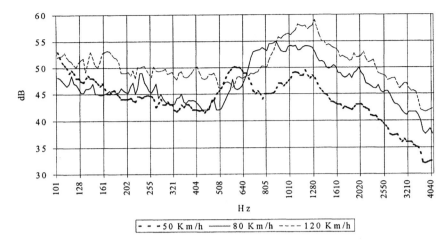

Figure 7: Spectral analysis of tire/road noise by low, medium and high speeds. (From reference [30].)

Tire/road noise is reduced at the source when porous roadways are employed. The main characteristic of such roadways is its greater **void's** ratio - typically > 20%. Noise emitted is smaller because the generation of so-called air pumping or air-resonant radiation noise is substantially reduced. Reductions from 1 to 8 dBA had been recorded in relation to conventional dense asphalt concrete surfaces. As a general rule porous roadways reduce noise emission above 500 Hz. On the other hand, the higher the speed, the greater the reduction. In order to evaluate acoustical performances measurements can be achieved using the recent ISO standard 11819-1 [31] which describes the so-called *Statistical pass-by method*. In this method, the maximum A-weighted sound pressure levels of a statistically significant number of individual vehicle pass-byes are measured at a specified road-site location along with the vehicle speed. Even though the obtained *Statistical Pass-By Index* (SPBI) is not a real equivalent level, it is based on similar calculation procedures in order to show the difference between roadways as they would occur if one had measured equivalent A-weighted levels. Results obtained for a new porous roadway [32] showed a remarkable diminution in the SPBI for the three speed ranges specified by the aforementioned ISO: 3.3 dBA for low speeds (45-64 km/ h); 4.3 dBA for medium speeds (65-99 km/ h) and 4.7 dBA for high speeds (>100 km/h).

The weak point of porous roadways is that after just one or a few years most of them become clogged and thus lose much of their acoustical efficiency. Such deterioration is slower in countries with warm climates. An improvement on this should be the feasibility of the periodic clearing of clogged porosities.

5.2 Railway noise: Characteristics

On an equal basis to vehicle noise there are four basic sources of noise in train operations: a) wheels/rail noise, b) engine noise, c) noise from the auxiliary equipment and d) aerodynamic noise.

Wheel/rail noise is most common and most often dominant on the railway. It arises because the wheel and rail are set into vibration by the action of rolling one over the other. The spectrum analysis of measurements reveals some response peaks emerging from a substantial content of broad-band noise, suggesting that both resonant-modes and broad-band response are being excited [33].

Roughness of the wheel tread and of the rail head are considered as the most relevant sources of wheel/rail noise and vibration, in the theories developed to explain railway noise [34], [35]. However, other sources have been identified which can be attributed to parametric excitation, including wheel imbalance, rail support by sleeper, ballast of varying stiffnesses, and variation of contact stiffness between wheel and rail. These sources may be relevant for broadband excitation [36]. In order to reduce wheel/rail noise the first actuation is to reduce the roughness of wheels and rails as much as possible. This has the effect of minimizing the varying components of force between wheels and rail and corresponds to using the best of current practice. Another option is to reduce the number of wheels per unit length of the train, such as the Spanish *Talgo* train.

In the last twenty years, and because train builders are pushed faster to enhance their competitive position, the maximum train speeds have vastly increased. Whereas wheel/rail noise increases roughly as the cube of speed, aerodynamic noise increases at around the sixth power of speed [37]. For speeds up to 250 km/h aerodynamic noise is noticeable even in the presence of wheel/rail noise. The high speeds of the *Shinkansen* trains of the *Japanese National Railways* have highlighted the appearance of new noise sources, which mostly originate in aerodynamic noise. Recent studies employed sound intensity techniques [38]. Thus for a 300 series car moving at 262 km/h different sources of noise have been quantified. With reference 1 pW, the sound power of different sources were as follows: bogie part: 109-117 dB; pantographs and covers: 108-110 dB; bus cable covers and coupling areas: 100-108 dB; the head and the rear part of the train: 100-112 dB. The three last sources can be dealt with as aerodynamic noise and its sound power levels are near to wheel/rail sound power.

5.3 Aircraft noise: Characteristics

From the 1960s onward, the noise associated with civil aircraft operations in the vicinity of airports has been one of the principal environmental problems in many countries. The beginning of turbojet aircraft operations notably increased the sound levels. In addition, a substantial growth in the number of flights has been taking place since then. The number of American people living in areas where the Ldn (day/night sound level) exceeded 65 dBA reached its highest point in the early 1970s. It is estimated to have been between 5 and 7.5 million people [39], [40]. The development of the high-bypass-ratio engines reduced noise pollution, particularly on takeoffs, even providing much higher propulsion efficiency. Fortunately the number of old and loud aircraft is decreasing permanently [41].

For a single event, usual descriptors in aircraft noise [42] are *Maximum Sound Level* (MaxL), *Perceived Noise Level* (PNL), *Sound Exposure Level* (SEL) and *Effective Perceived Level* (EPNL). For cumulative events, usual descriptors are *Equivalent Sound Level* (Leq), *Day/night Sound Level* (Ldn) and *Community Noise Equivalent Level* (CNEL).

The main source of noise during takeoff and landing operations is the aircraft engine. It comprises fan and compressor noise (internal sources) and jet noise. The dominant mechanism on fan noise generation is the rotor-stator interaction. Flow disturbances created by an upstream rotor interacting with stator vanes produces both tones and broadband noise. Broadband generation mechanisms are less well understood than tone generation mechanisms and are the subject of current research [43].

Jet noise is the most important aircraft noise source. There are two processes in the generation of noise. One is the shock structure of a supercritical pressure ratio jet (locally supersonic) and the other caused by the turbulent mixing process of the jet with the atmosphere.

6 Prediction of urban noise

6.1 Introduction

As was said at the beginning of this chapter, the prediction of noise is a powerful tool for several reasons. Generally the methods are used to assess the degree of exposure to noise for a given or projected road, highway, railway, airport, factory or any town planning capable of originating noise annoyance. The methods can be categorized into one of three groups. First, the manual methods permit the calculation of noise levels by using basic formulas together with tables or correction charts. The second type is a computer method. The powerful graphic programs used for showing results allow a clear display of them over the whole selected area like a noise map. Against that, an extremely large number of data about the sources and geometry are required for using this method. Finally, scale-model techniques are particularly suited for some complex specific situations. Because the present computer programs make use of standards models, our attention will be focused on such standards. Furthermore, since the most important source in urban noise is road traffic noise, the standards discussed here are those concerning road traffic noise.

6.2 Basic theory of traffic noise generation

For the prediction of noise due to road traffic, the most interesting type of source is a line of length L with a source strength W per unit length (see Fig. 8). Each segment dx can be considered as a point source with source strength Wdx [44].

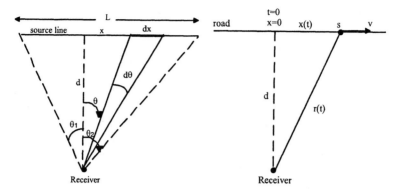

Figure 8: Geometry for a line source.

Figure 9: Geometry for a moving point source.

Let us suppose a flat and reflective surface as well as a source radiating equally in all directions, that is to say, the directivity factor $Q_{\theta,\phi}$ equals 2. Neglecting air absorption the intensity at distance d is given by:

$$I = \int_L \frac{W dx}{2\pi r^2} = \frac{W}{2\pi d} \int_{-\theta_1}^{\theta_2} d\theta \qquad (17)$$

If $L \gg d$ then θ_1 tends to $-\pi/2$ and θ_2 tends to $\pi/2$. In logarithmic form:

$$L_I = L_W - 3 - 10 \log d \qquad (18)$$

It can be seen that the level decreases by 3 dB when the distance is doubled for an infinite lineal source. When point sources radiate as incoherent dipoles another term must be included in the integral in eqn (17), e.g. for rail/wheel interaction noise from a train [45]. Let us suppose now a vehicle moving with speed V and emitting a constant sound power W by the road line (see Fig. 9). The sound pressure at the receiver is given by:

$$P^2(t) = \frac{\rho c W}{2\pi r^2(t)} \qquad (19)$$

Then the sound pressure level (for $P_{ref} = 20\ \mu Pa$ and $W_{ref} = 1\ pW$) can be written:

$$L_p(t) = L_W - 10 \log 2\pi r^2(t) \qquad (20)$$

For the minimum distance d, pressure received, P_d, is maximum and any sound pressure can be expressed in respect of P_d and distance:

$$P^2(t) = P_d^2 \frac{d^2}{V^2 t^2 + d^2} \tag{21}$$

taken t = 0 when vehicle passes by x = 0. For sound level:

$$L_p(t) = (L_p)_d - 10 \cdot \log\left[1 + \left(\frac{t}{\tau}\right)^2\right] \tag{22}$$

where $\tau = d/V$ and $(L_p)_d = 20 \log\left(\frac{P_d}{P_{ref}}\right)$

If a long time period is considered L_{eq} (1 h) can be easily evaluated. In this case t = T = 3600 s and the corresponding angle limits for the integration are $-\pi/2$ and $\pi/2$ (note that situations $V \cong 0$ and $d \rightarrow \infty$ have been neglected). Then the mean value for energy is given by:

$$\left[\overline{P}^2\right]_{-T/2}^{T/2} = \frac{1}{T}\int_{-T/2}^{T/2} P^2(t)dt = \frac{1}{T}P_d^2 \int_{-T/2}^{T/2}\frac{\tau^2}{t^2 + \tau^2}dt = \frac{1}{T}P_d^2 \cdot \tau \cdot \pi \tag{23}$$

where it has been assumed that T >> τ. In logarithmic form:

$$L_{eq}(1\,h) = (L_p)_d + 10 \cdot \log \tau + 10 \cdot \log \pi - 10 \cdot \log(3600) =$$
$$= (L_p)_d + 10 \cdot \log \tau - 30.6 \tag{24}$$

For a traffic flow of Q identical vehicles per hour, the L_{eq}(1 h) value obtained above is increased by the term 10 log Q. Then:

$$L_{eq}(1\,h) = (L_p)_d + 10 \log \tau + 10 \cdot \log Q - 30.6 \tag{25}$$

In terms of sound power level, according to eqn (20) and since $\tau = d/V$

$$L_{eq}(1\,h) = L_w - 10 \log d - 10 \log V + 10 \log Q - 38.6 \tag{26}$$

It may seem strange the dependence $L_{eq} - V$. If L_W is not dependent on V, the term $- 10 \log V$ means that when the vehicle moves at a higher speed, it remains a shorter time near the receiver and consequently the noise energy received decreases. Such is the situation for low speeds but for speeds above 60 km/h the sound power of vehicles depends on the third or the fourth power of speed.

It is clear that individual vehicles have different sound power. In the last equation L_W represented the average value for all vehicles concerning the same class, e.g. passenger cars, medium trucks or heavy trucks. A different sound power $(L_W)_i$ is assigned to each ith class to obtain the corresponding $(L_{eq})_i$. Total L_{eq} can be obtained by energetic summation of all $(L_{eq})_i$:

$$L_{eq} = 10 \log\left[\sum_i 10^{(L_{eq})_i/10} \right] \tag{27}$$

6.3 Standards

Prediction models for the calculation of traffic noise in the environment must be characterized in two parts: the definition of the noise source with its noise emission and source description and secondly the propagation of the noise from source to receiver. The general principles for the prediction of the noise level at a certain point are based on the propagation model. For the calculation of such a noise level a more explanatory equation than the correlative eqn (1) must be used:

$$L_{Aeq} = L - D_d - D_a - D_g - D_b + G_r - C_m \tag{28}$$

where:

L_{Aeq}: A-weighted equivalent sound pressure level at the receiver point
L: Noise emission of a source as a sound power level or as a sound pressure level at a reference point defined by the model
D_d: Attenuation due to geometrical spreading
D_a: Attenuation due to atmospheric absorption
D_g: Attenuation by ground effects
D_b: Barrier attenuation
G_r: Correction for reflections
C_m: Corrections for meteorological effects

Many countries have implemented standards for road noise prediction. Some of them are the following:

- Calculation of Road Traffic Noise (CRTN; United Kingdom)
- Richtlinien für den Lärmschutz and Straβen (RLS-90; Germany)
- Guide du Bruit des Transports Terrestres (GdB II; France)
- Beräkningsmodel för Vägtrafikbuller (BfV-89; Nordic countries)
- Reken-, en Meetvoorschrift Verkeerslawaai II (RMV II; The Netherlands)
- Überarbeitung des Rechenverfahrens des RVS 3.114 Lärmschutz (RVS 3.114; Austria)
- Federal Highway Model (FHWA; United States)

In addition there is ISO 9613 Part 2: General method of calculations, which provides formulas for the attenuation of sound emitted by point sources during propagation outdoors.

The majority of them predict the continuous equivalent sound level in broad band and for A-weighting, L_{Aeq}. The UK standard CRTN predicts the statistical descriptor L_{10} either for the loudest hour or for an 18-hour period. All standards share a basic setup. The sound power and directivity are not used as a noise description because of the lack of knowledge on acoustic emission characteristics of individual sources. Let us remember both the different contributions to noise emission in a vehicle-engine, transmission, tire/road contact, exhaust and their different spectra depending on the speed.

From the traffic and road characteristics models start calculating a basic noise level at a reference distance. The reference distance varies from 10 to 25 meters. Traffic and road characteristics affecting the source level are traffic density, percentage of heavy vehicles, speed for cars and trucks, type of asphalt and road gradient. The basic equation to calculate the basic noise level is in fact a way to assign a certain sound power per road unit length. Afterwards a propagation model evaluates other contributions such as distance to receiver, ground attenuation, viewing angle correction, screening, reflections and meteorological corrections.

Next we explain more about of three such models.

6.3.1 UK model (CRTN)

One of the oldest models for the prediction of traffic noise is called the Calculation of Road Traffic Noise. The following résumé is based upon the 1988 version. The basic noise levels to be calculated are the statistical descriptors L_{10} for a one-hour period and an average L_{10} for an 18-hour period. This basic level is calculated at a distance of reference of 10 m from the road corresponding to no heavy vehicles, a mean speed of 75 km/h, zero gradient and a conventional road surface:

$$L_{10}(1\ h) = 42.2 + 10 \log q \quad \text{dBA}$$
$$L_{10}(18\ h) = 29.1 + 10 \log Q \quad \text{dBA}$$

where q = total vehicle flow within the hour
Q = total vehicle flow, 06.00 to 24.00 hours

Correction for mean traffic speed and traffic composition:

$$\text{correction} = 33 \cdot \log\left(v + 40 + \frac{500}{v}\right) + 10 \cdot \log\left(1 + \frac{5p}{v}\right) - 68.8\ \text{dBA} \qquad (29)$$

where v = mean speed, km/h, during the relevant period
p = percentage of heavy vehicles in traffic flow during relevant period

Correction for gradient:

correction $= 0.3\,G$ when actual mean speed is used

where $G = $ percentage gradient

For low traffic volume and short distances to the source line there is a correction of the basic noise level:

$$\text{correction} = -16.6 \log D \text{ sqr}(\log (C)) \qquad (30)$$

where $D = 30/d$ with $d = $ shortest slant distance and $C = q/200$ or $Q/4000$.

Corrections for distances including types of ground are taken into account. Such corrections depend on the next values: slant distance between reception point and source line, height of reception point, ratio between them and percentage of absorbent ground.

Viewing angle correction is taken in the usual form: $10 \log(\theta/180)$, where θ is the angle view of the segment.

Correction by long barrier:
The insertion loss of a barrier is evaluated by a polynomial:

$$\text{correction} = A_0 + A_1 x + A_2 x^2 + ... + A_n x^n \quad \text{dBA} \qquad (31)$$

where x is the extra path length, $x = \log(a+b-c)$ (see Fig. 10). Values of constants $A_0, A_1,...,A_n$ are given in Table 2.

In situations where more than one barrier screens the reception point from the road, the screening corrections for each of the barriers should be determined separately.

With regard to reflections a correction of +2.5 dBA is required for reception positions 1 m from a building façade. If the opposite side of the road is completely reflective, the maximum correction due to reflections can be 1.5 dBA.

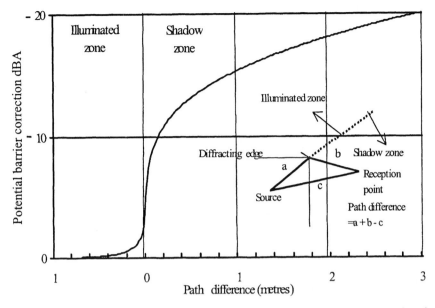

Figure 10: Potential barrier correction (very long barrier) as a function of path difference.

Table 2. Value of constants $A_0, A_1,...,A_n$

Constant	Shadow zone	Illuminated zone
A_0	−15.4	0
A_1	−8.26	0.109
A_2	−2.787	−0.815
A_3	−0.821	0.479
A_4	−0.198	0.3284
A_5	0.1539	0.04385
A_6	0.12248	
A_7	0.02175	
Range of validity	$-3 \leq \underline{x} \leq 1.2$	$-4 \leq \underline{x} \leq 0$

6.3.2 Federal Highway Model (FHWA)

The FHWA Highway Traffic Noise Prediction Model is another of the oldest models for calculating noise impact from road sources. It arrives at a predicted noise level through a series of adjustments to a reference sound level. The reference level is the energy mean emission level. The contributions from cars, medium and heavy trucks are considered separately. After that, adjustments are made taking into account traffic flows, distances from the roadways, the finite length of the roadways and shielding. All of those variables are related by the following equation:

$$L_{eq}(h)_i = (L_0)_{Ei} + 10 \log\left[(N_i * \pi * D_0)/(S_i * T)\right]$$

$$+ 10 \log(D_0 / D)^{1+\alpha} + 10 \log\left[\Psi\alpha(\Phi_1, \Phi_2)/\pi\right] + \Delta_S \qquad (32)$$

where

$L_{eq}(h)_i$ is the hourly equivalent sound level of the *i*th class of vehicles

$(L_0)_{Ei}$ is the reference energy mean emission level of the *i*th class of vehicles

N_i is the number of the *i*th class vehicles per hour

D is the perpendicular distance, in meters, from the center line of the traffic track to the receiver point.

D_0 is the reference distance, 15 meters in this model

S_i is the average speed of the *i*th class of vehicles, in km/h

T is the time period over which the L_{eq} is computed (1 hour)

α is a parameter depending upon site conditions

Ψ is an adjustment for finite length roadways

Δ_S is the attenuation, in dB, provided by barriers, wooded areas, etc.

Three classes of vehicles are dealt with in the model: cars, medium trucks and heavy trucks (see Table 3). Medium trucks have only two axles and the weight is between 4500 kg and 12000 kg.

The distance and absorption adjustments are taken into account by the term $10 \log(D_0/D)^{1+\alpha}$. When the ground is hard, $\alpha = 0$ and all the energy is reflected; when the ground is soft or covered with vegetation $\alpha = 0.5$. The model uses nomograms for the evaluation of adjustments for finite length roadways. The screening is based on the extra path length according to the Fresnel number theory.

Table 3. Types of vehicles

Vehicle type	Reference mean emission level $[(L_0)_{Ei}]$	Height of the source above the road surface
Passenger cars	$38.1*\log(S) - 2.4$	0 m
Medium trucks	$33.9*\log(S) + 16.4$	0.7 m
Heavy trucks	$24.6*\log(S) + 38.5$	2.44 m

6.3.3 German model (RLS-90)

From the mean hourly traffic volume (M, in veh/h) and percentage of trucks exceeding 2.8 tons, the basic noise level (at a distance of reference of 25 m and 4 m above the ground) is calculated by the equation:

$$L(25 \text{ m}) = 37.3 + 10 \log[M (1+0.082 P)] \qquad (33)$$

for the following conditions:

Speed of cars: 100 km/h
Speed of trucks: 80 km/h
Road surface with non-grooved asphalt
Gradient less than 5%

Adjustments for speed, type of asphalt, gradient and reflections are taken into account to evaluate the source level. The model deals with a range of speeds from 30 km/h to 130 km/h by cars and from 30 km/h to 80 km/h by trucks. The speed correction, C_S, is:

$$C_S = L_{car} - 37.3 + 10 \log \frac{100 + \left(10^{\,0.1 \cdot C}\right) P}{100 + 8.23 \, P} \qquad (34)$$

where:

$$C = L_{truck} - L_{car}$$
$$L_{truck} = 23.1 + 2.5 \log(V_{truck})$$
$$L_{car} = 27.8 + 10 \log[1 + (0.02 \, V_{car})^3]$$

For gradients higher than 5% a correction of 0.6 g − 3 is applied, g being the gradient of the road lane.

When walls are in opposition and multiple sound reflections occur between them the corrections, C_R, are the followings:

$$C_R = 4 \text{ H/L} \quad \text{(3.2 dBA at most) for hard surfaces}$$

$$C_R = 2 \text{ H/L} \quad \text{(1.6 dBA at most) for absorbent walls}$$

where:

H is the wall height
L is the distance between walls

The proximity of a traffic light increases the noise level. An addition of 1, 2 or 3 dBA may be added to the contributions of a road if the reception point is found at 100, 70 or 40 meters from a traffic light. The model assumes that both braking and accelerating operations increase the noise emitted at the same velocity. The propagation model evaluates corrections for finite length section, spreading and air absorption, ground attenuation and screening.

6.4 Comparison of the models

Leaving aside the different reference distances, there are differences in standards both at basic noise level calculation and at propagation model. One way in which the definition of the sound source differs in the different models is in their position on or over the road platform. Most models assume the source to be situated in the axis of the lane of traffic, with the exception of the UK model in which the source is situated at 3.5 meters from the near-side edge of the road. The height of the source is usually 0.5 meters- UK, Nordic, German and Austrian standards. For The Netherlands it is 0.75 meters and 0.8 meters in France. The USA model (FHWA) adopts three different heights depending on the type of vehicle: passenger cars (0 meters), medium trucks (0.7 meters) and heavy trucks (2.44 meters).

For the calculation of the basic noise level at the reference distance all models assume the dependence 10 log Q, where Q represents the density of the traffic. In most models, two types of vehicles are dealt with: light and heavy vehicles. In The Netherlands and USA models three categories are taken into account: light vehicles, light trucks and heavy trucks. The basic noise level at reference distance depends strongly on the composition of the traffic and its speed. For example, the German model assigns to heavy vehicles a sound power 9.2 times higher than that corresponding to light vehicles at a reference speed of 100 km/h. In the FHWA model and at the same speed, this relation is 12.6 for medium trucks in comparison to passenger cars and 24.5 times for heavy trucks in comparison to passenger cars. In all models, the basic noise level is increased with speed for light vehicles but in some models (Germany and The Netherlands) the basic noise level decreases with respect to the speed for heavy vehicles.

Concerning the propagation there are several differences among models. The first difference is basically the methodology concerning how a section of road is managed. Two points of view are general: angle method and point source method. The first one states that the noise level from a road decays with distance. The distance effect from a road is used to describe the entire road. If only one section of the road is to be calculated, the level from the section is described deducting 10 log(φ/180) from the level that the whole road would cause.

The second one states that a section of road is to be treated as a point source. All energy emitted from a section is concentrated at one point and propagated from there to the receiver point. The propagation does not require the perpendicular distance from the road to the receiver point. The influence of a section depends on 10 log(L) and a formula describing the spreading from the point source (Fig. 8).

The majority of models assume that the noise emission is towards a half space. Free fields models make use of the source emission as an omnidirectional source. For such models and for reflective ground the ground attenuation will be 0 dB. For the half space models the ground attenuation will be negative so the noise levels will be higher. As a general rule, the acoustic absorption of the ground is represented by a non-dimensional coefficient g with values between 0 and 1. g = 0 for hard ground such as asphalt, water or concrete. g = 1 for soft ground such as grass. For mixed grounds g is the percentage of the soft ground (0<g<1). The meteorological conditions can be both favorable and unfavorable for the propagation. Some models only apply favorable conditions for the propagation, that is to say, downwind or moderate temperature inversion. Other models make use of a correction for a typical average situation. Table 4 outlines [46] some of these comments for the above mentioned models.

In order to make a critical comparison of the different traffic noise calculations models, a separate study for the source and propagation model is appropriate because of the great number of variables taken into account. Moreover normalization in the reference distance is needed. Some results from a round robin test on road traffic noise schemes [47] showed remarkable differences.

Recalculating the source models to obtain the sound level per meter for the same traffic density of passenger cars, differences around 5 dBA were detected for different European country standards. The differences for heavy trucks are even larger: up to 10 dBA. These differences could not be accounted for by significant differences in the type of vehicles. Regarding the propagation models, similar differences were found for the excess attenuation due to screens. If all these effects would act in the same direction, total differences could add up to 10-15 dBA. At the sight of such differences it looks suitable to harmonize the standards. Because most propagation and screening methods already have much in common, harmonizing seems relatively easy. To harmonize source models the main difficulty lies in finding the right factors for the emission strengths.

Table 4. Outline of standards/models characteristics

Standard/ model	Type of sources		Frequency range		Space		Meteorological conditions	
	Point	Line	dBA	Octave	Free	Half	Average	Downwind
CRTN		•	•			•	•	
RLS-90	•	•	•			•		•
GdB II		•	•			•	•	
BfV-89		•	•			•	•	
RMV II	•			•	•		•	•
RVS 3.114	•	•	•			•	•	
FHWA	•		•		•		•	
ISO 9613	•			•	•		•	•

7 Methods for reducing urban noise

7.1 Introduction

Different actions can be taken in order to reduce urban noise mainly caused by road traffic. The first measure to consider is to reduce the emission in individual emission levels as well as limitations of flow traffic. Other actions point to either avoiding the propagation paths or absorbing sound. There is no doubt that the most efficient measure is the reduction of the acoustic emission from each vehicle. Technical improvements and more and more demanding requirements are resulting in a small reduction of noise levels in our cities [48], [49].

Reduction of 3 dB in the sound power from each vehicle yields the same effect as limiting traffic density by half. Obviously the last measure involves drastic consequences. Reduction on tire/road contact noise emission by using porous roadways was discussed in section 5.1.2.

7.2 Restriction of the noise emission levels

In the past 25 years numerous reductions in permissible noise values emitted by vehicles have been enacted. The European Union (EU) vehicle noise test procedure is listed in Council Directives 92/97/EEC and 81/334/EEC. For automatic-drive and high-power vehicles, this was amended to 84/372/EEC, based on ISO 362 [50]. This test is widely used throughout the world [51].

Table 5 lists the exterior limits that apply to vehicles sold in the EU. The evolution of vehicle noise regulations in the EU has been remarkable for the past 25 years. Table 6 outlines this evolution [52].

Table 5. European Union highway vehicle noise limits.

Vehicle class	Unloaded weight (ton)	Power (kw)	Limit (dBA)
Passenger car	------	------	74
Mini bus	< 2	------	76
	> 2 < 3.5	------	77
Bus	< 2	< 150	78
Light truck/ van	> 2 < 3.5	------	76
		------	77
Medium truck/ van	> 3.5	< 150	78
Heavy truck	>12	> 150	80
Motorcycles			
≤ 80 cc.	------	------	75
> 80 and ≤ 175 cc.	------	------	77
> 175 cc.	------	------	80

Vehicle noise regulations are more stringent in the EU than in the US [53]. With regard to vehicle noise-limit values in Japan, EU regulations are more stringent for passenger cars, buses and heavy trucks but less stringent for motorcycles.

Considering the fact that the evolution in vehicle noise regulations gives a diminution of about 10 dBA in noise levels, this is rather disappointing compared with the slight reduction observed in comparative studies carried out on urban noise evolution [49], [50], [54]. Table 6 outlines the evolution of the EU noise-limit values from 1972.

Obviously the increment in traffic density increases noise levels but it can not explain by itself the modest reductions. The reasons for the disappointing effectiveness in real traffic of the vehicle noise-limit values carried out in the last 25 years are manifold [55].

As regards the vehicles proper, the most important point is the major difference in the operating conditions if type testing – "worst case" is compared with real traffic situations. Likewise the test procedure should deal with a transmission ratio giving an acceleration as found in urban traffic.

Table 6. Evolution of the EU noise-limit values, in dBA, for the past 25 years

Vehicle class	1972	1980	1982	1989	1996
Passenger car	82	------	80	77	74
Bus	89	------	82	80	78
Heavy truck	91	------	88	84	80
Motorcycles					
≤ 80 cc.	------	78	------	77	75
>80 < 175 cc.	------	80 - 83	------	79	77
> 175 cc.	------	83 - 86	------	82	80

7.3 Reduction of noise levels on limiting the circulation of traffic

This limitation is normally applied to heavy vehicles in urban streets as well as in some roads during certain times of the day or week. It provides effectiveness due to the high acoustic power of such sources. In order to quantify the diminution in Leq values by limiting the circulation of heavy vehicles, let us suppose an acoustic power for them N times higher than the corresponding value for passenger cars, W_0.

If the total traffic density is Q (veh/ h) and the percentage of heavy traffic is P, the diminution in the value of Leq (ΔLeq) will amount to:

$$\Delta L_{eq} = 10 \log \frac{(1-P) \cdot Q \cdot W_o + P \cdot Q \cdot N \cdot W_o}{(1-P) \cdot Q \cdot W_o} = 10 \log \left(1 + \frac{P \cdot N}{1-P} \right) \quad (35)$$

Even though N depends on velocity a reasonable value for it is N = 10. If P = 0.1 (10% of heavy vehicles) the diminution in the value of Leq is 3.2 dBA. If P = 0.5 (50% of heavy vehicles) such diminution is 10.4 dBA.

8 Urban noise evolution in our cities

It is simple to answer the question: Has it reduced the noise in a given point of a city? For this, it would be sufficient to measure the sound levels at such a point both at the start and end of the required time interval. However, to answer that question for a city requires a noise descriptor characterizing the noise pollution for the city as a whole. Obviously, it also requires an identical methodology as well as the same scope of measurements at both times. Some comparative studies [48], [49], [54] show that the sound pollution in our cities is gradually reducing even though slightly. This occurs in spite of the fact that the traffic density has been increasing. Two circumstances seem be the most pertinent answers to explaining it:

a) The smaller sound emission of the individual vehicles.
b) The limitation of heavy traffic running into the cities.

The first measure is due both to the most exigent regulations as well as to the technical improvements in view of the greater demand for quiet vehicles.

As an example, for a wide measurement set carried out in a Spanish city [48], [56] in the years 1987 and 1997, a reduction of the environmental noise for the city as a whole has been given. The reduction of the "average" L_{Aeq} for the city was 1.4 dBA. This means a decrease of 27% for the sound pollution (in energy terms). Particularly important was the reduction of the L_{Amax} levels. Fig. 11 shows the cumulative distribution for both years.

With respect to the correlation between sound levels and traffic density Q (veh/h), Fig. 12 shows the dispersion diagrams for both years. The regression line Leq-logQ for both years seems to confirm a decrease in the acoustics emission of individual vehicles.

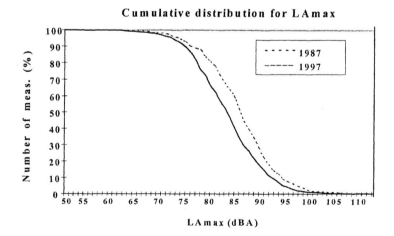

Figure 11: Cumulative distribution for L_{Amax}. Number of measurements: 648.

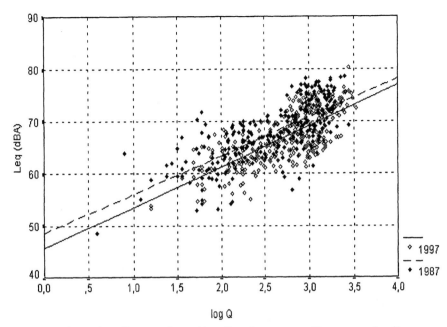

Figure 12: Dispersion diagrams Leq - Log Q and corresponding regression lines. (From reference [48].)

9 Final comments

As we have seen, factors affecting urban noise are multiple and very assorted. There are multiple types of sources with different characteristics both in spectral power and in directivity patterns. Likewise, factors affecting the spreading of sound have a notable influence in the sound levels at the reception points. The prediction of urban noise from the knowledge of the individual characteristics of the sources seems an impossible task in practice. Fortunately, the statistical treatment of the sources simplifies the problem. This is what we make when we consider a traffic road as a linear source with its acoustics power by length unit. This is actually what makes the standards and, consequently, the computational programs by means of assigning values to the influential variables of the traffic. However, we have also seen a notable contrast in the predictions from different standards. In this sense, it would be convenient to unify the standards, at least on the European level.

The determination through standardized tests of the sound levels emitted by vehicles in trial tracks can provide accurate values of the noise. Even the intensity techniques can evaluate the directivity diagrams of such sources. However, the sound emission of the same sources in real urban conditions is very different and not subject to standardization up till now. In this regard, the contrasting and notable decreases of the sound levels emitted by vehicles checked in such tests are not agreeing with the reduced decrease of the noise pollution detected in comparative studies in the last few years in various cities. Nevertheless the greater requirements for noise emission limits on vehicles may have produced the result that urban noise pollution has not increased, in spite of the increase of the traffic density.

In the last thirty years an immense amount of work in the identification and quantification of noise levels generated by various components of the environmental noise sources has been developed, especially relating to road traffic. To develop more noiseless components is the best means to control the noise in our cities. But it is essential to bear in mind that we humans are the ones who generate this noise pollution, either directly or indirectly through the use that we make of such sources. Because of this, environmental education is probably the best way forward in the struggle against noise.

References

[1]   Kinsler, L.E., Frey, A.R., Coppens, A.B. & Sanders, J.V., *Fundamentals of Acoustics*, 3rd ed., Wiley, New York, 1982.
[2]   Maling, G.C., Lang, W.W. & Beranek, L.L., Determination of Sound Power Levels and Directivity of Noise Sources, Chapter 4, *Noise & Vibration Control Engineering*, Beranek, L.L. & Ver, I. (Eds.), Wiley, New York, 1992.
[3]   Fahy, F.J., *Sound Intensity*, 2nd ed., E&FN Spon, London, 1995.
[4]   Crocker, M.J. & Jacobsen, F., Sound Intensity, *Encyclopedia of Acoustics*, ed. by Malcom J. Crocker. John Wiley & Sons, N.Y., 1997.
[5]   Reinhart, T.E. & Crocker, M.J., Source identification on a diesel engine using acoustic intensity measurements, *Noise Control Eng. J.*, **18**, pp. 84-92, 1982.
[6]   Attemborough, K., Review of ground effects on outdoor sound propagation from continuous broadband sources, *Appl. Acoust.*, **24**, pp. 289-319, 1988.
[7]   Embleton, T.F.W., Piercy, J.E. & Olson, N., Outdoor sound propagation over ground of finite impedance, *J. Acoust. Soc. Am.* **59**, pp. 267-277, 1976.
[8]   Stokes, *Trans. Cambridge Phil. Soc.*, **8**, p. 287, 1845.
[9]   Kirchhoff, G., *Pogg. Ann. Phys.*, **134**, 177. 1868.
[10] Herzfeld, K.F. & Litovitz, T.A., *Absorption and Dispersion of Ultrasonic Waves*, Academic Press, N.Y, 1959.
[11] Bass, H.E., Sutherland, L.C., Zuckerwar A.J, Blackstock, D.T. & Hester, D.M., Atmospheric absorption of sound: Further developments, *J. Acoust. Soc. Am.*, **97**, pp. 680-683, 1995.
[12] Sutherland, L.C. & Daigle, G.A., Atmospheric sound propagation, *Encyclopedia of Acoustics*, ed. by Malcolm J. Crocker. John Wiley & Sons, N.Y., 1997.
[13] Keller, J.B., Geometrical theory of diffraction, *J. Opt. Soc. Am.*, **52**, pp. 116-130, 1962.
[14] Maekawa, Z., Noise reduction by screens, *Appl. Acoust.*, **1**, pp. 157-173, 1968.

[15] Kurze, U. & Anderson, G.S., Sound attenuation by barriers, *Appl. Acoust.*, **4**, pp. 35-53, 1971.

[16] Pierce, A.D., Diffraction of sound around corners and over wide barriers, *J. Acoust. Soc. Am.*, **55**, pp. 941-955, 1974.

[17] Aylor, D., Noise reduction by vegetation and ground, *J. Acoust. Soc. Am.*, **51**, pp. 201-209, 1972.

[18] Embleton, T.F.W., Sound propagation in homogeneous deciduous and evergreen woods, *J. Acoust. Soc. Am.*, **35**, pp. 1119-1125, 1963.

[19] Verein, Deutscher Ingenieure, *Schallausbreitung im Freim*, VDI 2714, VDI-Verlag GmbH, Düsseldorf, 1988.

[20] Horoshenkov, K.V., Hothersall, D.C. & Attenborough K.. Porous materials for scale model experiments in outdoor sound propagation. *Journal of Sound & Vibration*), pp. 194, 685 - 708, 1996

[21] Horoshenkov, K.V., Hothersall D.C & Attenborough, K., Scale modelling of sound propagation in a city street canyon. *Journal of Sound & Vibration*, **223(5)**, pp. 795-819, 1999.

[22] Steven, H., Geräuschemission von Kraftfahrzeugen im realen Verkehr, *VDI-Berichte*, **499**, 9, 1983.

[23] Wilson Committee, *Noise - Final Report*, H.M. Stationery Office, London, 1963.

[24] Tyler, J.W., Sources of vehicle noise, in P.M. Nelson (Ed.), Chapter 7, *Transportation Noise Reference Book*, Butterworths, London, 1987.

[25] Lewis, P.T., The noise generated by single vehicles in freely flowing traffic, *Journal of Sound and Vibration*, **30**, p. 191, 1973.

[26] Priede, T., In Search of Origins of Engine Noise - an Historical Review, *SAE Congress*, Detroit, Paper N° 800534, 1980.

[27] Russel, M.F., Diesel Engine Noise: Control at Source, *SAE Congress*, Detroit, Paper N° 820238, 1982.

[28] Challen, B.J. & Morrison, D., Automotive Engine Noise. In Lilly, L.R.C. (ed.), *Diesel Engine Reference Book*, Butterworths, London, 1984.

[29] Lamure, C., Road Traffic Noise: Generation, Propagation and Control, in A. Lara Sáenz & R.W.B. Stephens (Eds), *Noise Pollution*, John Wiley & Sons Ltd, 1986, chap. 12.

[30] Barti, R., Tyre Noise Particularities, *Proc. of InterNoise97*, pp. 203-206, Budapest, 1997.

[31] ISO 11819-1, 1995, *Acoustics: Method for measuring the influence of road surfaces on traffic noise*. Part 1: Statistical pass-by method, ISO, Geneva, Switzerland.

[32] Sandberg, U., A new porous pavement with extended acoustical lifetime and useful even on low-speed roads, *Proc. of InterNoise97*, pp. 99-104, Budapest, 1997.

[33] Stanworth, C., Sources of Railway Noise, *Transportation Noise Reference Book*, P. M. Nelson (Ed.), Butterworths, London, 1987.

[34] Remington, P.J., The estimation of wheel/rail interaction forces due to roughness, *BBN Rep.* N° 7793, Oct. 1992.

[35] Thompson, D.J., Wheel - rail noise generation, part IV: Contact zone and results, *J. Sound Vib.* **161**(3), pp. 447-466, 1993.

[36] Kurze, U.J., Niederfrequente Schwingungen bei Eisenbahnen, *DAGA`92*, 337-340, 1992.

[37] Hickling, R. and Oswald, L.J., A Proposed Revision of the SAE J57a Tire Noise Testing Procedure, General Motors Research Laboratories, Warren, MI, Research Publication No. GMR 2816, 1978.

[38] Kawahara, M., Hotta, H., Hiroe, M. & Kaku, J., Source Identification and Prediction of Shinkasen noise by sound intensity, *Proc. of InterNoise97*, pp. 151-154, Budapest, 1997.

[39] Eldred, K.M., Model for Airport Noise Exposure on a National Basis:1960 to 2000, *Proc. InterNoise80*, pp. 803-808, 1980.

[40] EPA, Report to the President and Congress on Noise, *Senate Document Nº 92-63*, Feb. 1972.

[41] Kastka, J., Mau, U. & Siegmann, S., Standings and results of the research on aircraft noise - Longitudinal study at Düsseldorf airport 1987-1995, *Proc. InterNoise96*, pp. 305-310, Liverpool, UK, 1996.

[42] ANSI 512.9 - 1998, *American National Standard Quantities and Procedures for Description and Measurement of Environmental Sound*, Part 1, Acoustical Society of America, N.Y, 1998.

[43] Groeneweg, J.K., Advanced Subsonic Aircraft Engine Noise Reduction Research, *Proc. InterNoise96*, pp. 317-322, Liverpool, UK, 1996.

[44] Favre, B.M., Factors Affecting Traffic Noise, and Methods of Prediction, in P.M. Nelson (Ed.), *Transportation Noise Reference Book*, Butterworths, London, 1987.

[45] Peters, S., The prediction of railway noise profiles, *J. Sound Vib.*, **32**, pp. 87-99, 1974.

[46] Van Leeuwen, J.J.A. & Nota, R., Some Noise Propagation Models Used for the Prediction of Traffic Noise in the Environment, *Proc. of InterNoise97*, pp. 919-922, Budapest, 1997.

[47] Van der Berg, M. & Gerretsen, E., Comparison of Noise Calculation Models, *Proc. of InterNoise96*, pp.: 311-316, Liverpool, UK, 1996.

[48] Arana, M. & San Martín, M.L., Influence of car modernization on environmental noise in Pamplona, Spain. *Proc. 6th International Congress on Sound and Vibration*. Copenhagen, 1999.

[49] Kozák, J., Trends in Traffic Noise in Prague from 1976 to 1991, *Proc. of 17th AICB Congress*, pp. 61-66, Prague, 1992.

[50] ISO 362, 1961, *Measurement of Noise Emitted by Vehicles*.

[51] Morrison, D., Road Vehicle Noise Emission Legislation, Chapter 9, *Transportation Noise Reference Book*, ed. P.M. Nelson, Butterworths, London, 1987.

[52] *Green Paper*, ECC,COM (96) 540 final, Bruxelles, 1996.

[53] Hickling, R., Surface Transportation Noise, Chapter 88, *Encyclopedia of Acoustics*, ed. Malcolm J. Crocker. John Wiley & Sons. N.Y., pp. 1073-1082, 1997.

[54] Sandberg, U., Report by the International Institute of Noise Control

Engineering Working Party on the Effect of Regulations on Road Vehicle Noise, *Noise News International*, June 1995, pp. 82-113.

[55] Fingerhut, H.P., Requirements for an external Noise Measuring Procedure for Commercial Vehicles with Greater Environmental Effectiveness, *Proc. of InterNoise96*, pp. 3333 - 3336, Liverpool, UK, 1996.

[56] Arana, M. & Garcia, A., A social survey on the effects of environmental noise on the residents of Pamplona, Spain, *Applied Acoustics*, **53** (4), pp. 245-253, 1998.

Chapter 6

Urban noise control

Amando García
Department of Applied Physics, University of Valencia, Valencia, Spain

1 Introduction

The control of environmental noise is a subject of tremendous social and economic interest, which deserves the efforts of many professionals in fields as different as physics, engineering, architecture, urbanism, health, sociology, law, psychology, etc. Independently of which are the direct protagonists of the initiatives carried out to solve a specific problem, it should be stressed that the noise abatement in urban areas is a responsibility of the whole of society, namely the administration (in its different levels) and the citizens (as individual persons or as members of the corresponding communities).

Noise pollution is a most complicated subject, with different scientific, technical, economic, political and social aspects, with important repercussions for all society. Consequently, noise control requires the use of a large range of strategies. In a general sense, urban noise control should be based in a coordinate association of technical initiatives (careful evaluation of each specific problem and proposal of effective actions intended to solve it), an adequate legal framework (existence of regulations or normatives that orientates and endorses the corresponding actions), political determination (willingness to propose and apply decisions and continuity of all necessary actions) and social support (general recognition of the importance of this environmental problem and active contributions from all the members of a given community to the success of the proposed initiatives) [1][2][3][4][5].

In principle, modern technology is able to solve completely (or at least to alleviate considerably) most of the problems related with the environmental noise to which millions of people living in developed countries are exposed. Unfortunately, the economic cost or the social repercussions of the possible solutions limits or prevents their practical application in many cases. Certainly, noise control measures are always more effective and less costly when they are designed at a very early stage of development.

The formulation of an effective policy to control urban noise should be based on the techniques of planning, management and economy. Obviously, all these techniques rest upon a broad knowledge of the nature and characteristics of the existing problems and on the scientific or technical principles which define the possible solutions of such problems [6].

The articulation of an effective policy to control the environmental noise in urban areas involves a number of specific actions to be taken in a sequential order. The first phase of this process consists in a most careful evaluation of the arisen problems (identification of main noise sources, measurement of noise levels, etc.). In this sense, it should be remembered that the broad spatial and temporal variations of environmental noise levels in urban areas makes this task considerably more difficult (in general, this evaluation consists in the production of a detailed sound map of the considered zone). In particular, when new noise sources are to be introduced in a given urban area, it is necessary to evaluate the changes that these initiatives will produce in the present acoustical situation of such an area, through a study of environmental impact, based on a careful consideration of the before and after situations.

The second phase of the process should consist in a realistic formulation of the objectives to be reached in a given context (establishing the maximum values of environmental noise levels that should be permitted through the application of certain specific actuations). These objectives could be related with problems as diverse as the construction of a new motorway in the vicinity of a residential area, the enlargement of an existing airport, or the introduction in a big city of a new model of refuse disposal vehicles.

The third phase should be based in the application of the technical and administrative principles that could prove more effective in reaching the proposed objectives. Obviously, in the selection of the measures to be taken, in each case all their implications should be considered. For instance, when the construction of an acoustical barrier is projected in order to shield a residential area against the traffic noise coming from a nearby motorway, not only will it be necessary to evaluate its economic cost (which, in some cases, can be very high), but also its degree of acceptance by the affected residents (visual impact, aesthetic evaluation, etc.) or its possible effects on traffic safety.

The fourth and last phase of the process consists in the verification of the effectiveness of the adopted measures. If the objectives initially proposed are not reached it will be necessary to repeat all actions from the initial phase. Among many other factors, the possibility that a given noise problem can be produced by a number of different noise sources, and the fact that for each noise source and each receptor many different sound transmission paths generally exist can considerably impede the solution of many problems.

When designing possible initiatives to control environmental noise it should be taken into account that all environmental noise problems always consist of a chain with three different links: the noise source (with a specific power, spectra and directivity characteristics), the sound transmission path (through solid, liquid or gaseous matter) and the receiver (in the last instance, the people exposed to the noise). Consequently, the measures or actions taken to control noise can be applied only to one of these elements or to all of them conjointly.

In general, noise control on the source itself is the most appropriate option. For instance, regarding road traffic noise, these actions could consist in an improvement of the present vehicles' technology or an adequate control of the respective running conditions. The noise control on the source itself is especially advisable in the case of the noisiest vehicles (buses and trucks), given that their contribution to the general levels of traffic noise is usually quite important. In that sense, particular mention should also be made of the high noise levels produced by motorcycles, especially when they are running with faulty exhausts.

Regarding the noise transmission path from source to receptor (the second link of the above mentioned chain) the very important role played by the urban-style conditions on the noise abatement in urban areas should be emphasized. Many of the initiatives proposed to control the environmental noise pollution in cities are based on perceptive urban planning. These initiatives are related to measures such as increasing the distance between noise sources and the noise-sensitive zones (through open spaces, plantings, etc.), interposing some noise-compatible activities such as commercial facilities or parking spaces between the noise sources (a motorway, a noisy industry, etc.) and the most noise-sensitive activities (residences, schools, hospitals, etc.), or using some form of noise barriers to shield the specially sensitive areas.

Finally, with reference to the reception of noise, the best example of noise control is the use of personal protection devices against noise by workers exposed to high occupational noise levels in the working place (in a similar sense, the same situation could correspond to the acoustical insulation of buildings in order to minimize the immission of external noise). There are two basic types of hearing protection devices: earplugs and earmuffs. The earplugs are generally made of soft plastic or silicone rubber and are inserted into the ear canal. Earmuffs consist basically of hard plastic cups with a cushion seal which cover the outer ear as an acoustical shield or barrier. When properly used, these devices provide a high degree of sound attenuation.

2 Road traffic noise control

Road traffic (trucks, buses, passenger cars and motorcycles) is, by far, the most important noise source in both urban and suburban settings of all developed countries. Actually, road traffic noise disturbs more people than all other forms of noise nuisance combined [7]. In the last decades, traffic noise exposure has increased considerably as a result of the combination of growth in urbanization and increased mobility by the population. It has been estimated that about 20% of the population in developed countries are currently exposed to road traffic noise equivalent sound levels Leq exceeding 65 dBA, measured outside of the building façades [2]. In these conditions, the sleep and conversation of residents can be significantly disturbed even if the corresponding dwelling windows are closed. Consequently, the control of traffic noise should be considered a priority in all general plans of environmental noise abatement in urban areas.

There are three main methods of controlling the impact of traffic noise on communities. The first approach is to attempt to reduce noise at its source by the design of quieter vehicles. The removal of vehicles (reduction of traffic flow), reduction of traffic speed and re-routing strategies should also be included in this category. The second approach involves different attempts to limit the spread of noise, once it has been generated (control of the sound transmission path). In that sense, it has been observed that the road design has significant effects on the traffic noise levels. The use of low noise special road surfaces has also produced important reductions in the noise pollution levels. Initiatives such as the construction of acoustical barriers or an adequate planning of the land use alongside a main road can also contribute to reducing the noise disturbance for adjacent residents. The third approach refers to the use of noise protective measures at the receiver. An example of this last strategy refers to the acoustic insulation of new or existing buildings to minimize the immission of road traffic noise in their most sensitive zones. Walls, windows, doors and roofs are the main elements of sound insulation in buildings and, generally, the quality of these components determines the degree of insulation achieved by building façades as a whole. Which method, or which combination of methods are actually employed in each case depends basically on the degree and nature of the noise reduction required and upon the influence of both economic and operational constraints [4][6].

In this Section the main strategies of traffic noise control will be briefly revised: quiet vehicle development, road traffic management, road design, special road surfaces, land use planning and acoustical barriers. Acoustical insulation of buildings will be treated in a later Section.

2.1 Quiet vehicle development

Of the different road traffic noise abatement strategies available, the reduction of noise at the source is the most obvious [8]. However, the achievement of significant reductions of the noise emitted by vehicles requires substantial investments in research. Since the benefits of these actions are intended basically for society as a whole, the manufacturers have in the past been quite reluctant to invest in developing quieter vehicles.

It is necessary, therefore, that the relevant administrations (generally, at an international or national level) establish some incentives to encourage the required technological innovations. These incentives generally take the form of legislative actions establishing noise limits for both new and in-service vehicles. In an attempt to create a market sensitive to vehicle noise, a wide variety of forms of economic instruments have also been proposed.

To develop methods of controlling the noise produced by different types of road transportation vehicles, it is necessary to understand how the various vehicle components generate noise, how these components are designed and how the noise emitted by these sources changes with the vehicle operating conditions. The main sources of road vehicle noise have been identified as the power unit (engine, air intake and exhaust), cooling system, transmission (gearbox and rear axle), rolling noise (aerodynamic noise and tire/road interaction), brakes, pumps for water, fuel and lubricating oil, and a number of ancillary systems. The relative importance of these sound sources depends on the quality of vehicle maintenance and the operating conditions. Because the individual noise sources in a vehicle combine logarithmically to produce overall vehicle noise, it is imperative that all the individual sources are reduced together since little improvement is obtained if, for example, engine noise is reduced by about 30% and exhaust noise is left untreated [3][9].

The improved engineering and design of all these components can produce a substantial reduction in vehicle noise emission. For instance, the engine noise reduction may be achieved by reducing the magnitude of the combustion and mechanical forces in the acoustically significant range of frequencies, modifying the transmission paths of these forces by introducing additional damping into the engine structure, or diminishing the engine sound radiation efficiency by using appropriate enclosures or shields. In particular, the exhaust noise (produced by the sudden release of gas into the exhaust system when the exhaust valve opens) may be substantially reduced by installing in the vehicles adequately designed exhaust systems [8][10].

The total sound power emanating from a vehicle is minuscule compared to the power involved in the operation of the vehicle. For example, the sound power of a passenger car in usual conditions is of the order of several milliwatts, whereas the power involved in operating the vehicle is of the order of kilowatts. This fact could, perhaps, explain why for many years noise reduction has not been considered a factor of prime importance in the design of road vehicles. However, since 1970 several countries have initiated some research and development programs aimed at producing a future generation of road vehicles with substantially reduced noise level emissions. The most important vehicle quietening programs have been on commercial vehicles (concentrating on engines, exhaust systems and cooling systems) and buses (complete engine encapsulation for rear and mid engined urban buses is now a quite common practice). In general, the objectives of all these projects have been to determine the technical feasibility of vehicle quietening by providing noise control engineering solutions for individual vehicle types.

For instance, the program research carried out on this subject in France involved the main car manufacturers (Renault, Peugeot, etc.) and a number of research laboratories. The expenses were provided both by the car industry and Government administrations. The objectives of the program concerned identification and assessment of main vehicle noise sources, noise reduction of the different components, evaluation of quietened vehicles and development of special sound-proofing materials. The primary aim was to provide industry with general expertise regarding noise reduction, rather than to quieten a particular type of vehicle to reach a given target. Nevertheless, some prototype vehicles were also developed. To mention only two examples, in 1976 Renault company developed a delivery lorry with an emission noise level of 84 dBA instead 89 dBA of the original model, and shortly afterwards an urban bus was provided with a rear engine with a noise level of 80 dBA instead of the prior 90 dBA [8].

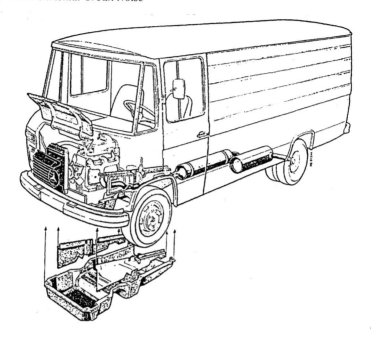

Figure 1: Example of a noise reduced delivery truck [11].

A number of vehicle noise reduction programs have been carried out in Germany starting in 1976. The programs involved several Government departments (Federal Ministry of the Interior, Ministry of Research and Technology, Berlin Environmental Agency, etc.) together with the main car manufacturers (Audi, Daimler-Benz, Volkswagen, BMW, etc.). As early as 1978, Daimler-Benz developed a private car with an emission noise level of 74 dBA instead of 79 dBA of the prototype model. However, considering that heavy vehicles are on average about 10 dBA noisier than cars, the main effort has been devoted to developing low-noise lorries and buses. Some German cities (following the example of Bad Reichenhall) have designed specific programs for the promotion of low-noise vehicles. The significant reduction of the noise level achieved in these vehicles is above all a result of encapsulation of the engine (Figure 1). The price of these vehicles is about 5% higher than the corresponding version without noise insulation [8][11].

The original legislation governing sound levels for motor vehicles (cars, lorries and buses) was adopted by the European Commission in 1970 (Directive 70/157/CEE). The latest amendment of this Directive came into force in 1997 (Directive 92/97/EEC). As the noise emission limits have been reduced, the tire/road noise has become more significant. The future regulations on this subject should contemplate carefully the relationship between noise reduction and economic costs before more stringent limits are established [5].

It should also be noticed that, due to the strong increase of traffic volume in all developed countries, the reduction of noise emission of motor vehicles these last decades has not produced a real reduction of the road traffic mean noise levels. If mobility continues to increase in the same way over the next few years, then the outlook is not very favorable, since the technical possibilities of a further noise reduction in vehicles are at present quite limited [12][13].

2.2 Road traffic management

The most important factors affecting the noise generated by road traffic as a whole are the traffic volume, the composition of vehicles in the traffic flow, the traffic speed and the characteristics of traffic flow. Road traffic is usually described as free flowing in motorways, and as interrupted flowing in most urban areas, where the presence of traffic lights and junctions produces many stop-and-go situations and, quite frequently, serious congestion [14].

An obvious way to reduce traffic noise consists in moving the traffic far away from the noise sensitive receivers (buildings) located close to the road. However, absolute closing of certain urban roads to traffic can present important problems of access. Some cities have established less severe traffic restrictions in certain urban zones, allowing normal access to such zones to public transportation vehicles (buses, taxis, etc.) or to private cars of the area's residents (who are provided with identification cards).

The effect of traffic volume controls depends mainly on the proportion of vehicles removed from the stream. Halving the traffic flow, for example, will generally lead to reductions in Leq noise levels of the order of 2-3 dBA. However, traffic volume and speed are generally highly correlated and so a reduction in volume is normally associated with an increase in traffic speed with the result that the optimum benefits expected from the reduced flow are not achieved. Furthermore, removing traffic from one road produces an increase in noise on adjacent roads in the network. However, the fact that traffic noise level and traffic flow are logarithmically related can be used to obtain some positive effects; for example, transferring traffic from a lightly used road and placing it on an already heavily used road will generally produce little additional noise burden on the heavily used road, but the benefits achieved on the lightly used road can be very substantial. Consequently, the construction of road by-passes, specifically designed to absorb high traffic flows from a network of residential streets, can produce a very important decrease in the noise annoyance levels of large numbers of people. Nevertheless, this measure is very costly and sometimes leads to considerable protest by the communities affected (urban residents or farmers). In any case, the construction of road by-passes requires very careful and far-reaching planning [15].

Strategies used to reduce the noise produced by road traffic also include restricting the number of heavy trucks in traffic flowing in urban areas. These techniques take usually the form of prohibitions on such vehicles entering a prescribed district, either in the form of a total ban on all commercial vehicles above a certain capacity (generally excluding buses) or in the form of restrictions at certain times, usually at night. In that sense, it is sometimes possible to divert heavy traffic along less sensitive routes (e.g. through industrial zones). The amount of noise reduction that can be achieved depends, of course, on the amount of heavy traffic that can be diverted. For instance, in a typical town situation with 15% heavy traffic a total diversion would lead to a noise reduction of 6-7 dBA. If 50% of heavy traffic remained, the noise reduction would be only 1-2 dBA [8]. The city of Berne, Federal Capital of Switzerland, probably provides one of the best examples of a combination of traffic measures to reduce urban noise levels, including a ban on trucks, vehicle-free zones and a very quiet public transport system [16].

Theoretically, the reduction of traffic speed is one of the most effective measures to control traffic noise levels. On high-speed roads (where rolling noise is predominant), halving the average vehicle speed could lead to Leq noise level reductions of about 9 dBA [17]. However, such reductions of vehicle speed cannot be easily achieved in practice. Speed limits are commonly imposed in urban areas (for instance, the general speed limit existing in all towns in Spain is at present 50 km/h), but actually a large proportion of motorists exceed these limits. It has been shown that the introduction of speed limits as low as 30 km/h in residential areas, as long as they are accompanied by certain traffic restriction measures, produce a reduction in the noise level of passing vehicles of up to 6 dBA, with no indication of an increase in exhaust emissions [11]. Various traffic calming schemes have been introduced in some countries with the aim of reducing vehicle speeds, the most effective usually involving the use of road humps or speed cushions across the road surfaces. Although important noise reductions can easily be obtained with this strategy (amounting to 5-7 dBA where the traffic consists only of cars), it has been observed that as the percentage of commercial vehicles increases the noise reductions become progressively smaller [18]. Other techniques used sometimes to reduce speed of vehicles include road narrowing and road bending.

An adequate design of the traffic speed restriction method is essential. The measures taken should introduce sufficient restraint on the motorist to produce speed changes without affecting gear changing, which could result in a net increase in noise levels. The methods adopted should also ensure that traffic flows freely through the site to encourage a non-aggressive style of driving. It should be noticed that speed control measures can have other positive advantages. For example, in the UK it was found that painting stripes on road surfaces on the approach to roundabouts (to give the drivers a feeling of high velocity) reduced accidents in the vicinity of those junctions by 50% [15].

The noise emission from road vehicles can increase substantially during acceleration, especially when the initial speed is low [18]. Vehicle acceleration and deceleration occur usually on the approach to junctions. Therefore, in order to reduce noise, it is important to consider, in the design of the junction, how to smooth the flow of traffic to minimize the number of vehicle accelerations. This objective coincides with the traffic management plans designed primarily to reduce journey times and accidents.

Linked or demand-controlled traffic light systems have been developed and installed in practically all large cities in the world. Unfortunately, the effect of these measures on traffic noise is less than expected, partly because the improvements in flow resulting from these control systems tend to generate an increase in capacity of the system which is rapidly filled by more vehicles; on the other hand, these systems also favor an increase of the mean speed of traffic. In general, the reduction of noise levels following the installation of a traffic-light control system is about 2 dBA. Another measure which has been used to smooth the flow through junctions is to switch off the traffic lights at low density junctions during the night. Such measures, however, have not given rise to any systematic improvement in noise levels since vehicle speeds are generally increased, which offsets the advantages arising from the fewer incidences where the vehicles accelerate from the rest.

In general, the roundabouts produce fewer noise problems than signalized intersections. Sound level measurements carried out at many roundabouts have indicated that the increased noise from accelerating vehicles is within 1 dBA of the free flow level on the approach roads and that noise from the decelerating stream is equal or less than the free flow level [15].

Once again, the reduction of road traffic, and most especially the private vehicle traffic, seems to be a key factor in reducing the urban traffic noise levels. It has been estimated that, in urban traffic, the private cars transport on average only 1.1 occupants, 50% of all car journeys are shorter than 3 km, and 25% of journeys under 1 km are made by car. Considering these and other data, it is thought possible that perhaps as much as 30% of the present number of private car journeys in many towns could be easily transferred to other modes of movement (public, bicycle or pedestrian transport) if these were extensively improved and encouraged. In that sense, the situation existing in many towns of Holland and the Nordic countries could be an excellent example [11].

2.3 Road design

The noise produced by traffic can be influenced by the vertical and horizontal alignment of the road. Roads can be elevated above the surrounding land at grade or can be depressed in cuttings. Elevated roads generally produce greater environmental noise problems but some screening will occur for reception points located below the edge of the embankment or parapet. Roads in cuttings are generally well screened by the cutting wall although reflections from the far wall can reduce the screening performance.

Depths of road cuttings can be very variable, depending mainly on terrain conditions. For the most usual depths (from 3 to 10 meters), it has been shown that noise levels are not greatly affected by the depth of the cutting because the improved screening provided by the increased depth of cut is offset by the increase in reflected noise from the opposite wall of the cut area. Where it exists, the separating wall at the center of the cut screens the direct sound from the second traffic lane causing substantially reduced noise levels in the vicinity of the edge of the cutting. Figure 2 shows an example of the noise field surrounding a section of depressed road with a depth of 5 meters. Improvement in the screening provided by this type of configuration can be obtained by reducing the reflectivity of the surface of the reflecting wall or by sloping the reflecting wall away from the vertical. Where the space is not limited, the cutting can be formed by embankments with slopes in excess of 45 degrees. The noise reductions achieved are in the range of 6-8 dBA [15].

The design of differently cut structures is usually evolved by using scale models. It has been found that absorptive lining placed on the retaining walls generally resulted in an additional noise reduction of about 3 dBA within 25 m from the edge of the cutting and up to 6 dBA at greater distances. In that sense, tunnels and enclosures form a natural extension to the retained cut form. These structures offer the advantages of very substantial noise screening over conventionally retained cut structures, particularly for high-rise buildings located close to the road. The disadvantage is the much higher cost involved and the need to provide artificial lighting and good ventilation. A further problem is related to the fact that the noise emission levels at tunnel openings can be very high, since the sound generated inside tunnels does not dissipate but rather reverberates within the structure. On the other hand, the gradient increase at portal ramps increases the noise emission levels produced by heavy vehicles as they accelerate and emerge from the portals [20].

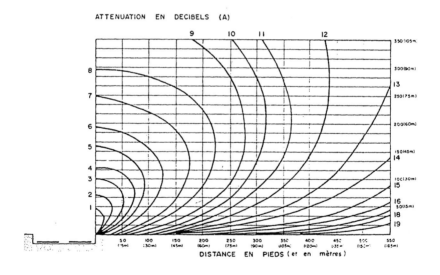

Fig. 2: Isophon curves obtained by simulation of a depressed road (sound attenuation in dBA versus distance in feet and meters) [19].

2.4 Special road surfaces

The noise generated by the action of tires rolling on road surfaces can represent a considerable contribution to the noise levels emitted by moving vehicles. Actually, it has been shown that the tire/road noise is predominant for cars moving at speeds above 50 km/h and for heavy vehicles at speeds above 80 km/h. In free-flow conditions (100-130 km/h) the tire/road interaction accounts for about 90% of total noise produced by cars [21]. Many studies of tire/road noise have established that while some benefits can be obtained by appropriate design of the tire tread pattern and tire structure, the design of substantially quieter tires conflicts with the need to maintain safety, cool running and economy. Consequently, the main efforts for reducing the tire/road noise have been devoted to developing special road surfaces. Over the past two decades, the development of special road surfaces has become a most important factor in the quieter road design plans [15][22].

The road surface characteristics which appear to be most important in determining surface noise are the texture applied to the road surface and whether the surface is a bituminous material with random texture pattern or a concrete surface with a predominantly transverse texture. Some studies have shown that for tires running on different surfaces, the road surface influence is largely dependent upon the macrotexture of the surface. In new conditions, effects of 1-8 dBA of noise reduction in relation to conventional dense asphalt or concrete surfaces have been recorded. The high dispersion of results is due to design and production differences in the compared surfaces.

Measurements carried out in several countries have established a simple empirical relation between the skidding resistance afforded by the surface and the total noise generated by vehicles passing over this surface at high speeds. Although this result is useful in setting standards for road surface finish which take into account both safety and environmental considerations, it highlights the existing conflict between the specifications of low-noise surfaces and adequate high-speed safety standards. A very smooth road surface can be relatively quiet but is clearly unsafe for the motorists, especially in wet weather.

Fortunately, some road surfaces do not conform to this general pattern and appear to offer the combined advantage of both low noise and good skidding resistance performance. The pervious macadam material was found to retain its noise reduction properties during the effective life of the surface material and, due to its rapid drainage properties, it lessens the incidence of that so called splash noise during wet weather. Similar results to that obtained with pervious macadam have been obtained with an open graded covering using a rubberized asphalt as binder. A test surface laid over a grooved concrete section of road on the ring motorway east of Brussels reduced the noise levels by about 4 dBA at car passing speeds of 70 km/hr and 5.5 dBA at a speed of 120 km/hr. Similarly, in Canada an open graded carpet seal was found to reduce traffic noise levels. It was shown that this material is about 4-5 dBA quieter than a worn concrete surface which was far less skid resistant [15].

In countries with a hard winter, the road surface performance deteriorates rapidly. It has been observed that, after just one or a few years, these pavements become clogged and thus lose much of their acoustical characteristics. Cleaning processes to reduce the clogging have been applied but have not been very successful so far. At present, many efforts are being devoted to the design of new, efficient and durable pavements [23][24][25].

2.5 Land use planning

A major road or motorway always generates high noise levels in its immediate vicinity. When a new traffic route is planned through an urban area much of the existing buildings will remain unchanged. Under these circumstances the design of the road becomes crucial in minimizing the noise impact resulting from the traffic. When the road passes through an undeveloped or scheduled for redevelopment urban area, the appropriate management of the adjoining land use can also be considered. Strategies for successful acoustical planning are determined by the size of the available space and the zoning policy applied. Appropriate strategies are based on actions such as placing as much distance as possible between the road and the particular activity, placing noise-compatible activities between the road and the site in question or using walls and plantings as barriers to screen sensitive areas [15][26].

In particular, dwellings can be protected from traffic noise by setting them well back from the source of noise (major road or motorway). However, this simple approach often fails to receive proper attention from the designers because it is assumed to be uneconomic. It should be noticed, for example, that on a level location next to a busy motorway the noise level will rarely be below 70 dBA at distances less than 100 meters from the road. Nevertheless, spatial separation should always be considered for in certain circumstances it is the only possible way to solve a noise problem. This is especially true in mixed urban developments which include high-rise blocks, for these cannot easily be screened by barriers and should, therefore, be located as far as possible from the road. On such sites, the remainder of low-rise dwellings can often be protected by some form of roadside barrier.

Some local administrations apply special policies to the land adjoining any major road, restricting the future development of these zones to non-noise sensitive activities (such as warehouses or certain industries). Unfortunately, in many urban areas, there is not enough demand for such noise-compatible land use to afford adequate protection for every community exposed to noise. Furthermore, this type of land zoning may not be compatible with other plans for the orderly growth and development of a given urban community, or it could be in direct conflict with the development plans of the adjacent communities.

In many cases, the use of buildings as noise barriers is a very interesting possibility. A long building or a row of buildings parallel to a highway can shield other more distant structures or open areas from the noise. In a study carried out some years ago in the four-storied building of the Faculty of Physics of the University of Valencia (located close to a busy motorway) it was found that the noise level on the side of the building away from the road was about 12 dBA lower than on the exposed façade [27].

In that sense, cluster developments enable the whole space to be planned as a single entity taking into account the use of both space and noise-compatible structures as noise buffers. Figure 3 shows an example of a cluster development concept in the vicinity of a major road. The placement of some light industrial or commercial buildings near to the road can provide some screening to the closest dwellings. Of course, the dwellings located farthest from the road are not as well screened as above, but benefit instead from the increased distance. Consequently, an acceptable acoustical environment can be achieved for all residents in the community [15].

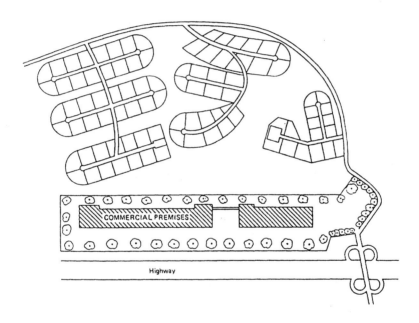

Fig. 3: Example of a cluster development in the vicinity of a major road [12].

2.6 Acoustical barriers

Application of acoustic barriers to highways has become the most common procedure for the abatement of road traffic noise [15][29]. In general, the roadside barriers may be considered to form part of the road design. An acoustical barrier is normally some form of vertical wall, although a wide range of designs have been adopted in practice mainly to attempt to improve the aesthetics of the noise screen rather than to improve its screening performance. A successful noise barrier must possess sufficient mass to attenuate the sound, it must be relatively maintenance free once installed and it must not produce an increased risk of accident. Other objectives are that it should be economical to construct and have an acceptable visual appearance [28].

In order to provide an optimum degree of protection, the barrier should be erected near to the noise source or close to the position to be protected and, if possible, it should completely obscure the view of the road from the buildings or space to be protected. As illustrated in Figure 4, the screening provided by an acoustical barrier depends upon the path difference between the shortest path over the top of the barrier between the noise source and receiver and the length of the direct line between these two points. Consequently, the effectiveness of a barrier increases with its height [29][30].

It should also be noticed that the sound energy generated by road traffic can be reflected by a barrier wall negatively affecting the receivers located on the source side of the barrier. When there are barriers on both sides of the road a further problem may occur as a result of multiple sound reflections between the two barrier walls. With these configurations the screening potential of each barrier can be significantly reduced as a result of additional noise diffracted over the barrier from the image sources.

In the United Kingdom the height of most road barriers is limited to 3 meters, whereas in the USA, Canada and some other European countries, higher barriers are permitted. Anyway, barriers higher than 4 meters are generally considered to be visually unacceptable to residents. Apart from height, the shape of the barrier is also important. A simple wall is generally less effective than a rampart of the same height. Several scale-model investigations have shown that T-profile barriers produce greater noise attenuation than equivalent conventional barriers. Some authors have also suggested mounting some type of absorptive material on the top of the barriers [31].

Wind loading, snow loading and the possibility of ice forming in the shadow of a barrier are all important safety considerations to be considered. The materials used should also withstand vehicle impacts. The construction of barriers over structures such as bridges or viaducts need special considerations because of the need to safeguard roads passing underneath from falling debris resulting from a vehicle impact. Obviously, an effective barrier should not deteriorate rapidly under the action of weather agents.

Many different types of acoustical materials have been used in barriers. Briefly, barriers can be classified as either reflective or absorptive. The reflective type is the more common (prefabricated panels). Close-boarded wooden barriers are fairly typical but other materials include vinyl plastic, pre-cast concrete, earth barriers, cellular concrete, aluminium and caged rock type constructions. Absorbing barriers are intended to reduce the amount of reflected sound from the barrier surface. Again many designs have been suggested and tried. A typical construction consists of a hollow box type panel which, on the motorway side, has a perforated or open steel face. The box is then filled with an absorbing material such as mineral wool.

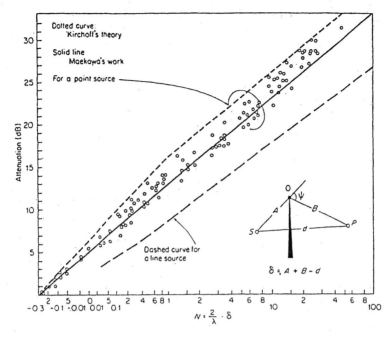

Fig. 4: Attenuation provided by a semi-infinite noise barrier [30].

Everybody recognizes that trees, bushes and plants are of great value in improving the aesthetics of a roadway environment. However, traffic noise attenuation provided by vegetation is generally overestimated. Vegetation may affect the propagation of low-frequency sound by ground absorption, which can be enhanced in wooded areas because of the high porosity of the ground resulting from the tree roots, fallen leaves, etc. High-frequency propagation is affected by scattering by tree trunks, branches, and partly by leaf absorption. Nevertheless, it is very difficult to provide precise descriptions of vegetation which can be used to gauge the attenuation of noise. Tree height, depth of planting, and particularly density of planting (kg/m^3) appear to be the dominant variables. Excess attenuation (with respect to bare earth) in the range of 0.5 to 2.0 dBA per 10 meters of dense and evergreen vegetation depth have been reported. In spite of these modest figures, some authors suggest that, under certain conditions, vegetative screens could provide an acceptable reduction in road noise levels at a fraction of the cost of a conventional barrier [32].

3 Aircraft noise control

Airports are a most important component of the worldwide transportation network. During the last few decades the number of operations involving civil and military movements has been growing rapidly, including both fixed (planes) and rotary-wing (helicopter) aircraft and both using civilian (public or private) and military facilities. In the USA, for example, there are nearly 20,000 civil and military airports and it is estimated that aircraft transport generates over 100 million flights a year. Current predictions are forecasting a doubling of the number of aircraft passengers within the next decade in all developed countries. This increase in air traffic, coupled with the congestion which already affects many major airports, is providing the impetus for accelerated plans of airport constructions, and the utilization of the latest technology to improve airport and airspace capacity. However, aircraft noise is a major impediment to the growth of existing airports and the construction of new ones.

The main mechanism of noise generation in the early turbojet aircraft was the turbulence created by the jet exhaust mixing with the surrounding air. The new generation of jet aircraft with high by-pass ratio turbofan engines (which surround the high velocity jet exhaust with a lower velocity airflow generated by the fan) has produced a substantial reduction in the noise impact caused by commercial air traffic in the vicinity of large airports [33].

The aircraft noise limits established by different administrative institutions have become increasingly restrictive over time. For instance, according to US Federal Regulations, the oldest and loudest Stage 1 commercial aircraft (for example, Boeing 707-100) are no longer permitted to operate in the United States, while Stage 2 aircraft (Boeing 727-200) were not allowed to operate after December 1999. Consequently, this restriction only leaves the newest and quietest Stage 3 aircraft (Boeing 757-200) in service [34].

While the control of noise impact generated by aircraft can be considered to be primarily the concern of the aircraft engineer to develop quieter aircraft, clearly this can be shown to be only a part of the solution. Aircraft noise control has to be also extended to the management of the complex interactions of airports and their surrounding communities. In that sense, there are three basic approaches to controlling the airport impact on neighboring areas: controls for the flying aircraft (air control), controls for airport general activity related both with aircraft operations and services (ground controls) and land use controls for the area surrounding the airports (application of regulations and economic incentives) [35][36].

3.1 Operational air controls

The noise control operational methods require the cooperation of the air traffic managers, the operating airline and the airport authorities. The noise control measures include a wide range of possibilities: establishment of preferential runways, aircraft noise monitoring, curfews, noise abatement flight tracks, enacting of operational fees, etc. [37][38].

Designing a preferential runway system is a very important strategy in use in many large airports. When possible, this procedure can produce an important reduction of the noise exposure over the more populated runways. However, it should be noted that, in most cases, runway assignments depend on local weather conditions. Of course, at small airports, which have a single runway system, the application of this technique is not possible.

In general, flight tracks are prescribed for individual departures and arrivals in order to avoid overflight of noise-sensitive urban areas. The flight track is the projection onto the ground plane of the three-dimensional flight path of the aircraft. There is a flight track for both approach and departure operations. For noise abatement purposes, the flight track is useful for positioning the aircraft in space relative to ground or land uses. Accordingly, many airports have assigned headings that place aircraft over non-populated areas (water, agricultural land, uncultivated areas, etc.). Once the flight tracks have been optimized at a specific airport, opportunities for reduction of noise exposure may exist by establishing the appropriate utilization of these corridors.

A variant of the natural flight track dispersion is the use of fanning, a deliberately controlled program for allocating the flight tracks. There are a variety of techniques employed to apply this system. The practice of fanning appears to be considered acceptable when the noise associated with each track is below the threshold value of significant annoyance for the exposed population. However, if the resulting noise is greater than this threshold, fanning tends to increase the number of people negatively affected [39].

The manner in which the aircraft operate during take-off (departures) or landing (arrivals) is a key factor in determining the noise levels around the airports. Figure 6.5 illustrates a departure profile consisting of three distinct phases. The objective of the phase 2 is to limit the noise exposure for people living in the vicinity of the airport during the take-off operations until the aircraft reaches sufficient altitude. It should be observed, however, that the reduction in the rate of climb can worsen the situation for people living quite far away from the airport. Of course, when selecting a specific alternative, the first requirement is ensuring the safety of aircraft.

The term "curfew" applies to the time that aircraft are generally permitted to operate in a given airport. The time restrictions are considered by the authorities and air companies as the most stringent form of aircraft noise control. Although the curfews can have significant economic consequences for the air companies (especially when air transportation involves multiple time zones), many airports throughout the world have established some form of curfews. For example, Switzerland's International Airport at Geneva imposes a complete night-time operational curfew between 22.00 and 06.00 hours for all aircraft traffic. Boston Logan International Airport prohibits scheduled departures of noisier aircraft during the night. London's Heathrow International Airport permits only a limited number of operations at night during the summer period.

Noise monitoring on a continuous 24-hour basis by permanently installed equipment is the most viable prospect. Consequently, the interest of airport authorities in acquiring these facilities has been growing these last years. At present, there are about sixty of such systems operating in twenty different countries throughout the world. Airport noise monitoring systems consist basically of four components, including a series of remote monitoring stations, a central processing station, computer hardware and software and some graphic map terminals. Most of these systems provide a graphic display of the monitored noise levels, often in real time. Such noise monitoring systems serve a wide variety of purposes, depending on the needs of the airport authorities: assessment of alternative flight procedures for noise control, investigation of public inquiries or complaints, assistance in addressing land-use planning, detection of unusual flight events, etc. The main reason for the use of these systems in most European countries is the enforcement of the corresponding mandatory noise level regulations [34][39].

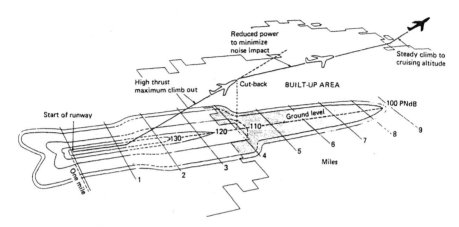

Fig. 5: Departure profile used to minimize the impact of aircraft noise [34].

3.2 Airport ground controls

Management of an airport can have a significant influence in controlling noise. The importance of the airport noise impact on the neighboring community is determined by factors such as operational running procedures, internal and external surface transport networks, service building locations or existence of noise shields or noise barriers. Application of certain effective administrative controls can also be a very positive initiative [34].

Most large airports maintain facilities for the maintenance and repair of aircraft. As part of this process, aircraft must undergo static tests requiring a certain thrust of engine power settings. Auxiliary power units and other auxiliary equipment may also be further noise sources in an airport. Since most of this run-up activity occurs during the low utilization periods (at night or early in the morning), these ground operations can produce an important noise impact on the nearest urban communities. In these circumstances, enforcement is usually achieved through specific local ordinances.

3.3 Land use controls

Improvements in aircraft technology are not in themselves sufficient to adequately control aircraft noise. Therefore, land use planning is vital as a most effective way to minimizing noise exposure of residents around airports. Incompatible land development is an undesirable by-product of improper planning, especially when it adversely affects the health and welfare of many people.

There are many land use control methods available to ensure that noise compatible land planning occurs in the vicinity of airports. Comprehensive plans, zoning ordinances, land acquisition and fitting dwellings with sound insulation are some of the possibilities to consider. Although the application of these land use measures is international, European countries have been the most aggressive in implementing these provisions [34].

Basic elements of a comprehensive plan, which should be intended to guide the long-term growth of an area, include the private uses of land (residential, services, commercial and industrial areas), community facilities (educational and health centers, sports and leisure settings, recreational parks, shopping centers, etc.) and transportation systems (roads, railways, underground lines, etc.). All of these elements involve decisions influencing land use compatibility and potential environmental impact. Among many other factors, the recognition of a wide variety of community generated noise is an important ingredient in a successful and comprehensive plan. The use of a general plan for local or regional development should reflect existing and future airport related needs.

Several European countries have developed governmental regulations for airport noise planning. For instance, in the middle of the 1980's, the State of North Rhine-Westphalia (Germany) prepared a Regional Development Plan, applicable to several civil and military airports. This plan established that in the so-called Noise Protection Area A (noise levels exceeding 75 dBA), there would be applied a general ban on creating new development plans containing localities which are habitually sensitive to noise (residential areas, hospitals, homes for the aged, educational centers, etc.). In the Noise Protection Area B (noise levels in the range from 67 to 75 dBA) the same ban applied as above, except for the rounding off of existing residential areas; in this case, sound insulation constructional measures would have to be taken, if necessary. Finally, in the Noise Protection Area C (noise levels in the range from 62 to 67 dBA), the plan applied a proviso to balance both regional and land use planning aims when making up new development plans.

In general, these comprehensive plans are implemented through zoning ordinances. As an effective legal tool, a zoning ordinance should regulate various aspects of land use development, including the height, bulk and density of buildings, area of land which may be occupied, size of required open spaces, density of population, and permitted uses of structures. These uses are usually specified for each district, including the relevant physical requirements. For instance, when the new Munich airport (Germany) was projected, the zoning regulations established that no new housing zones could be planned in the area with Leq > 67 dBA. In between Leq = 62 and 67 dBA, new houses could be built only in a limited number [40].

Besides these zoning requirements, most regulations also prescribe the minimum standards for the construction of dwellings in the vicinity of airports. These building codes are designed to guarantee the health, safety and welfare of the affected community. Regulations of this type have not been widely applied in the United States. In contrast, most European cities and particularly those around large airports do have ordinances with noise provisions. Many of these regulations have been enacted at the national level. All Dutch airports (both civilian and military) have noise zone boundaries: depending upon the noise contour boundary, noise abatement must be in compliance with some specific sound attenuation. Similar noise zones are established at all German airports. Up to date, over 135 million Deutsche Marks have been spent on implementing the sound insulation requirements at 47 airports in Germany. In the United Kingdom, 10 public airport authorities have been implementing a long-term plan in order to insulate over 50,000 public and private dwellings. The cost of this plan up to 1990 exceeded 40 million pounds sterling; the Government contribution ranged from 65-100%, depending upon the dwelling's location and noise exposure level. In recent years, the majority of houses around Copenhagen airport have been insulated, with state subsidies of 50-90%. Several studies of sleep quality of residents before and after insulation have shown that noise insulation results in considerably improved sleep quality. However, it should be noticed that building insulation represents only a partial solution to protecting property from aircraft noise since the corresponding outdoor environment remains unaffected [34].

4 Railway noise control

The different types and models of trains (short and long distance trains, passenger and freight trains, diesel or electric trains, low and high velocity trains, metropolitan trains, tramways, funicular railways, etc.) constitute important sources of noise which affect large geographical areas around the world. Considering that most trains pass across urban zones, and that many stations are situated in city centers, the impact of the noise generated by this transportation system adversely affects people living near rail tracks. Train noise pollution can result in complaints, conflicts and lawsuits, and it can produce even a loss of public support for rail transportation as an alternative to automobiles. Consequently, the control of railway noise should be an important component of any generalized plan of urban noise control. Implementation of available control techniques will decrease the adverse noise impact produced by train operations without sacrificing the ultimate goal of transporting many people and goods in a safe, timely and efficient manner [41].

There are many methods of controlling noise produced by trains. Most of them are similar to those used on road noise control (development of quieter vehicles, land use initiatives, etc.). In a more specific sense, the reduction of groundborne noise and vibration produced by trains is achievable by reducing the wheel/rail interaction forces and by attenuating the propagation of noise and vibration. The corresponding treatments can consist of controlling the emission at the source, use of a resilient track support structure, construction of tunnels and screening barriers, and application of relevant building segregation techniques. In some cases, these treatments are part of the original design of the railway system. In others, they are applied on a retrofit basis, in an attempt to alleviate some particularly serious noise problem [42].

4.1 Noise from train operations

The wayside noise radiated from train activities is the most important aspect of train noise. The noise and vibration produced in different train operations can take different paths depending on the type of support structure. Airborne noise is radiated from at-grade and elevated operations, while ground-borne noise is of primary concern for subway operations. Wayside noise emission is the result of a number of different factors including the locomotive propulsion system (traction motors, cooling fans and reduction gears), the interaction of the wheels and rails (noise radiated directly from the vibrating wheels and rails), auxiliary equipment (compressors, motor generators, brakes, ventilation systems, etc.) and noise radiated from elevated structures (vibration of the transit structure components excited by a train pass-by) [41].

Each of these noise sources can dominate the wayside noise level depending on operating conditions. For instance, at low speeds (less than 20 km/hr), the noise produced by auxiliary equipment predominates. At moderate speeds (about 50-100 km/hr), the prevailing noise comes from wheel/rail interaction. Of course, the above figures should be considered only illustrative, provided that the response of the corresponding systems depends on the specific conditions existing in each case [42].

Engine exhaust is usually the main noise source of diesel locomotives. Noise radiated directly from engine bodies may be also very important. In some designs, the engine is bolted directly to the locomotive chassis. Consequently, the engine crankcase vibrations pass into the external vehicle structure without having been attenuated by the engine mounts. The possible modification of the locomotive casing, in order to provide an adequate barrier to the engine noise, presents many problems (space constraints, need to provide adequate ventilation of oil fumes and easy access for maintenance, among others).

Excitation of the locomotive structure by engine vibration is best tackled by using resilient engine mounts. However, to make resilient connections to the engine for air, fuel, coolant, lubricant and possibly exhaust ducting is not an easy task. In any case, the possible solutions must guarantee that the electrical, hydraulic or mechanical power take-off not be affected [43].

The control of wheel/rail noise is achieved applying some specialized techniques such as removal of wheel surface irregularities, rail grinding, rail welding and rail alignment. In particular, it has been observed that wheel deformations, bad rail joints, or corrugated rails can increase the wayside noise levels by as much as 10 dBA compared with smooth wheels running on smooth continuous welded rails. In that sense, it may be expected that any arrangement that reduces the severity of impacts which occur as wheels traverse rail joints contribute to a reduction of the associated noise and vibration. Replacement of jointed rail with continuous welded rail is most beneficial, since welded joints eliminate all rail discontinuities. Careful maintenance and alignment of bolted joints is also desirable in the tracks where such joints are present [43].

The replacement of standard wheels with resilient ones diminishes the stresses between the wheel and the rail. The resiliently treated wheels reduce rolling noise through an increase of the contact surface and a decrease of its stiffness. The stiffness of rail fastener system, especially the elastomeric rail pad usually inserted between the rails and the sleepers, has a significant influence on the noise emitted by the track (Figure 6). Railways are increasingly using softer pads to reduce potential damage to sleeper and ballast. Softer pads decouple the rail from the sleeper. This uncoupling reduces the noise from the sleeper but also reduces the decay of vibration with distance along the rail and hence leads to an increase of the noise radiated by the rail [44][45].

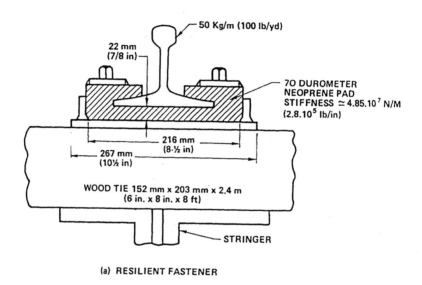

(a) RESILIENT FASTENER

Figure 6: An example of a rail low-noise fastening system [43].

4.2 Noise produced in stations

Another aspect of the problem refers to noise produced in stations. There are four main noise sources in railway stations: (a) trains entering, leaving or passing through the stations, and coupling and uncoupling of train coaches, (b) maintenance and storage yards, electric substations, and warning bells and horns, (c) a large series of ancillary equipment, such as ventilation, heating, and air conditioning systems, and (d) road transportation vehicles, shouting people, and public address systems, among many others [41].

The noise produced in a railway station can be annoying both for the public using its facilities and for people living in the neighborhood. It should be noticed, for instance, that the noise level at a station platform during a train's arrival or exit can be very high (perhaps more than 100 dBA) if the station lacks sufficient noise control measures. In many cases, the immission levels of noise produced by a station on the nearest dwellings considerably exceed that produced by road traffic or other community sources.

The principal strategies of controlling noise in transit stations consist of the application of adequate sound absorption measures in the busiest buildings, a careful design of ancillary equipment and maintenance machinery to meet sound emission criteria, and the insertion of acoustical barriers or some service buildings to reduce the impact of the most important noise sources on areas close to the stations.

4.3 Bridge noise control

From the early times of railway transportation it was generally recognized that trains passing over bridges could generate high levels of noise. A steel bridge forms a large, relatively lightly damped structure, which in some cases may also rattle when subject to vibration input. It is a fairly common practice to mount the rails on waybeams directly connected to the bridge structure, so that vibration from passing trains is very strongly coupled to the bridge. In such conditions, the wayside noise level may be as much as 20 dBA above that which would arise from trains at the same speed on track laid on the ground [43].

Most railway bridge construction types have several weak points. The problems are usually related to the rail fixation (where there is not enough resilience or damping) or the framework sited just under the rail (it may not be stiff enough or there may be too much radiating surface). Consequently, a logical way to reduce the input power into the bridge is to increase the resilience between rail and bridge construction. The easiest way to achieve this is to replace the rail pad with a softer one (cork and rubber pads are frequently used). It should be noticed, however, that once the pads have been replaced, the rolling noise produced by wheel and rail interaction increases. This problem can be avoided through the application of rail dampers [44].

Many European railways have been working on the reduction of railway bridge noise over the last 15 years. Some of the standard solutions consist in building steel bridges with a ballast bed, or building concrete bridges, either with or without ballast. All three of these bridge types imply high construction heights (distance between bottom of bridge and top of rail), a condition which, in some cases, can increase considerably the cost of the bridges.

The use of acoustical barriers (screening) is a well-established technique for controlling airborne train noise. A large variety of barrier configurations has been built. Both low barriers (close to the track) and high barriers (further away from the track) have been constructed. This technique can be applied both to ground level and to elevated level sections of the tracks. Absorptive and non-absorptive models of barriers have been used [46][47].

5 Community noise control

The different transportation vehicles (road traffic, aircraft and railways) are no doubt the most important and generalized environmental noise sources. Almost without exception, all people living in developed countries are exposed to the noise produced by some of these sources. Nevertheless, the residents in urban areas are also frequently exposed to many other noise sources that, depending of the conditions, can also produce serious annoyance for many people.

The term community noise refers usually to the noise produced by a wide variety of sources (vehicles, machines, equipment and industries) intended to cover the general needs of a given community. For instance, the noise produced by refuse collection vehicles, street cleaning vehicles or warning sirens and alarms, both stationary and mobile, should also be considered as community noise. The shops and markets, the public works and building construction, the workshops and small industries located in urban areas or the leisure and recreational activities (sports stadiums, fun fairs, discotheques, etc.), among many others, are other examples of community noise sources.

A third category of urban noise sources contributes to the so-called neighborhood noise. A wide variety of dwelling and home appliances (lifts, air conditioners, washing machines, dish-washers, telephone rings, door bells, radio and television sets, etc.) are usually included in this category. Some gardening chores (for instance the use of lawn mowers), noisy home hobbies, shouting and voices of people, bangs of doors, weeping of children or dogs barking, are other possibilities in this endless list. Obviously, all these noise sources can disturb both the residents of the dwelling where they are or the people living in neighboring dwellings. In many cases, the best way to reduce this disturbance is by a conscious social behavior.

In this Section, only a limited (but quite varied) selection of the large catalogue of community noise sources will be considered, namely the refuse collection vehicles, the public works and construction, and the leisure activities. In each case, the main characteristics of the respective noise source and the most significant techniques applied to control the noise produced by all these activities will be considered.

5.1 Refuse collection vehicles

Compression-type rear loader vehicles form the bulk of municipal refuse collection fleets in many developed countries. The size and power of these special vehicles exhibit a wide variability. The mechanism used to compact waste can work intermittently or continuously. Obviously, all these operational features determine the emission noise levels (typically in the range from 75 to 85 dBA) and, consequently, the extent of noise disturbance experienced by the residents of the cities where such vehicles operate.

An obvious approach to reducing the noise from refuse collection vehicles consists in developing quieter vehicles, to keep them in the best maintenance conditions, and to operate them properly (for instance, avoiding unnecessary strokes on the refuse containers). A substantial amelioration of the noise problems produced by the refuse collection can be achieved by scheduling collection times that do not disturb the sleep of residents. In certain urban areas, however, collecting refuse during working hours can be very difficult due to problems of traffic congestion. Where feasible, the possibility of enacting legislation to confine the refuse collection times to before 23.00 hours and after

07.00 hours should be considered. These times are actually observed in many cities of the United Kingdom with few complaints [48].

5.2 Public works and construction

The noise produced by public works and construction constitutes a frequent nuisance for urban residents. The risk of hearing damage in construction workers exposed to high occupational noise levels for 8 hours per day and over many years of their lives is also a serious problem. For this reason, the construction machinery and equipment industry has made substantial efforts over the last few decades to reduce the corresponding noise emissions.

Of course, the attitude of urban residents towards the public works and building noise is subjective and it can be positive or negative depending on many factors (evaluation of the object under construction, understanding of the necessity of the works, clear and adequate information about the intensity and duration of noise emissions from the works, acceptance of the planned working hours, etc.). It should also be remembered that public works are a temporary noise source, quite different to permanent noise sources such as road traffic, railways, aircraft or industries, and therefore much easier to tolerate [49].

Most of the construction machines produced and used at the present in European countries must comply with very restrictive sound emission regulations. In 1979, the European Community published the Commission Directive 79/113/EEC specifying the test procedures for construction machinery. The earlier sound pressure measurements were soon replaced by sound power controls that describe the sound emission of a given machine independently from the receiver distance (microphone position). This procedure is based on work done previously by the International Standard Organization (ISO).

In 1984 the framework Directive 84/532/EEC was published, requesting EEC approval for sound emission on all construction equipment. Yet in 1984 a number of special Directives addressing specific construction equipment were published: driven compressors (84/533/EEC), tower cranes (84/534/EEC), welding generators (84/535/EEC), generator sets (84/536/EEC) and concrete breakers (84/537/EEC). Two years later a new Council Directive appeared, addressed to hydraulic excavators, rope-operated excavators, bulldozers, loaders and backhoe loaders, setting noise limits that all these machines should accomplish to gain approval (86/662/EEC). An authorized test agency must provide a compliance certificate indicating a guaranteed sound power level for each individual unit of a type. In the last instance, the manufacturer is held responsible for ensuring that not a single machine is sold into the public work domain which exceeds the certified and labeled emission values [5].

In order to stimulate markets to demand construction equipment having sound emission levels lower than required by Directive 86/662/EEC, Germany introduced a few years ago a category of "Environmental Friendly Equipment". A machine will be classified in this category and will carry a special distinctive mark only if an authorized sound test office has certified its compliance with the established emission limits. The German administration is encouraging the machine users (loaders, compressors, generator sets, excavators, etc.) to buy equipment classified in this low noise category, especially when the construction sites are located in noise sensitive zones [49].

In practice, there are many technical solutions to reducing the noise impact produced by a building yard (using different machinery and equipment) on the most immediate neighborhood (shops, hotels, dwellings, etc.). The erection of a noise barrier around the building yard site is one of the most efficient. With careful design of this barrier a noise level reduction of as high as 20 dBA can be obtained [50].

5.3 Leisure and recreational activities

Many efforts have been made lately to fight against occupational noise exposure, and the situation in many countries concerning noise preventive measures seems promising. Unfortunately, the situation is much worse with regard to noise exposure in recreational activities. In that field, the number of noisy sources is very large: loud toys, pop and rock concerts, discotheques, car stereos, personal cassette players, racing motor vehicles, house and garden hobbies, hunting and competition weapons, etc. It has been shown that, depending on specific conditions, the leisure and recreational activities can generate important harmful effects [51]. As early as 1974, the Noise Advisory Council, working on the United Kingdom Noise Abatement Act (1960), issued a Report on Noise in Public Places that discussed many aspects of the noise problems related with public places [52]. According to this report, this category should cover all places to which the public have free or paid access, such as parks and gardens, public open spaces, beaches, sports stadiums or licensed premises. From a methodological point of view, the analysis of this problem should consider both noise that disturbs local residents and noise generated in any public place that disturbs people outside.

At the present, most of the rules controlling noise in public places are contained in noise regulations or Government bylaws promoted by local authorities after a consideration of the specific conditions and circumstances related to a given activity or noise problem. Consequently, the extent to which these regulations have been adopted by local authorities varies considerably from one country to another and even from one city to another within a given country. The consistency and effectiveness of the initiatives and actions promoted to solving a given problem can also vary considerably depending on many technical, social or political factors.

For instance, in the case of open-air activities such as organized sporting events, fun fairs or pop/rock festivals, the corresponding planning authority can take a number of steps to prevent the creation of a noise nuisance on the people living in the neighborhood of such activities. It should be noticed that different motor sports (car racing, rallies, motorcycle trials, etc.) are probably the noisiest of all sporting activities; with few exceptions, most of these activities are performed far away from densely populated residential areas. Unfortunately, this is not the case with many stadiums, which are frequently located inside cities, and that in addition to scheduled or unscheduled sporting events, are also used for other purposes (pop or rock concerts) characterized by a significant sound emission. In order to avoid a generalized adverse public reaction in the nearby community, a number of noise control measures have been proposed or applied in these cases: precise public information on the programmed activities, monitoring of sound emissions to guarantee that the permissible noise limits are not exceeded, control of other environmental or social problems related with the activity (parking, littering, vandalism, drugs, etc.), limitations to the duration of the spectacles, control of the public address systems, alterations in the existing structures, installation of sound insulation dispositives, etc. [53][54][55].

A possible way of preventing the noise nuisance produced by some leisure activities can be grounded on land use planning techniques. However, it should be noted that to control the noise produced by some existing developments of this kind is not an easy task, as, in some instances, the owners or occupiers of the land have acquired a right to use them long before the adequate planning controls could be introduced. In these cases, only the situations related to an intensification of use or change of use could be regulated. In certain cases, the local authority can give permission for the use of land for sports, social events, fun fairs or other purposes for a limited time, for instance, 14 or 28 days in a calendar year. In these cases, the authority should establish all the conditions under which temporary permission is conceded. As a part of these conditions, the limits of noise emission should be clearly established [52].

In recent years, an increasing source of complaints in many countries has been related to discotheques, bars, dancing clubs, youth associations and other similar entertainment facilities. The problem arises because of the high levels of amplified music used in most of these activities which, in many cases, continue into the early hours of the morning. It should be also noticed that, as with other urban noise sources (airports, industries, etc.), the sound impact produced by these activities is both direct and indirect. In other words, the noise annoyance experienced by near residents is not only related to the use of high level amplified music and to the voices of a large number of people filling a given space (closed or open-air), but also to the road traffic and crowd movement (people arriving or leaving these premises). The direct sound impact of the activity (noise produced inside the premises) on people living in the more or less immediate neighborhood can be reduced by a restriction on the opening hours and/or by an adequate sound insulation of the premises concerned; both aspects can be easily regulated by the corresponding local administration [56].

In a recent study on the noise impact produced by bars, pubs and discotheques, carried out by our University in the Community of Valencia (Spain), the equivalent sound levels Leq measured inside the premises ranged from 70 to 115 dBA, with peak levels as high as 130-140 dBA. For young people who are frequent visitors to discotheques or pubs, or for employees and disc jockeys, the exposure to these levels represents a real risk of hearing damage. In this study, it was found that for people living close to these places (sometimes even in the same building), the sleep disturbance produced by noise coming from the premises or streets where they are located, can be very important, especially during the weekends, where the official closing time of these premises is 02.00 or 03.00 hours, depending on their characteristics. A better administrative regulation of these particular leisure activities is revealed as an urgent necessity [57].

6 Acoustical insulation of buildings

In urban environments, the reduction of noise immission by the building itself should be considered as the last resort in urban noise control strategies, to be applied only when there is insufficient reduction at the source or in the path between the source and building. The sound insulation of building sections (walls, floors, ceilings, etc.) is determined by the physical properties of the material, the type of construction and the method of installation, and varies with the frequency of incident sound.

It should be remembered that sound can be transmitted through building construction as a result of either airborne or structure-borne paths [58]. The basic approaches to control airborne sound transmission include the selection of materials that reduce the degree of transmission, the elimination of paths of direct air transmission, and the use of sound absorbing materials within the free airspace of frame construction. Often a combination of these control measures is necessary to achieve acceptable sound levels inside buildings. Structure-borne sound transmission results when the different surfaces of the construction are set into vibration through acoustical coupling or direct mechanical contact. Structure-borne transmission comes from sources such as footfalls, dropped objects, doors slamming or induced vibrations from mechanical systems and equipment (underground, heavy vehicles, lifts, etc.). Main methods of control include the increase of transmission loss of walls, ceilings and floors, and the use of discontinuous construction assemblies [59][60].

The need for expensive construction with high sound insulation can be minimized if the shape and orientation of the building is planned considering the path of noise coming from the road. The aim should be to avoid reflected sound from any wall surface being directed into the noise-sensitive rooms of the building itself or any nearby building. The shape of the building can be utilized to provide a self-protecting structure where some parts, such as wing walls or balconies, provide shielding from road traffic noise. For balconies located well above the level street, the cover should be designed so as to reflect sound away from windows (which generally have low sound insulation) or it should be furnished with a sound absorbing material.

Within any building there are some rooms in which the people are less annoyed by noise from outside than in other rooms. As road traffic noise is usually only a problem for the rooms facing directly towards the road, the more noise-sensitive rooms should be located on the other side of the building. The less noise-sensitive rooms (kitchens and toilets) can then provide a barrier to the penetration of external noise into the more noise-sensitive zones (living rooms and bedrooms) in residential buildings. For existing buildings a similar approach can be used to minimize the need for expensive actions to guarantee that immission noise levels do not exceed suitable values. For instance, in large commercial buildings there are usually a number of different types of rooms and uses, so a rearrangement of the internal space usage can allow for the noise-sensitive areas to be away from the road façade [15].

Methods of construction and materials used in buildings vary widely both within and between countries [3]. All these changes have an important effect on the sound insulation of the buildings. The following Sections give a brief overview on the general performances of the basic elements which form the building enclosure.

6.1 Wall acoustical insulation

The sound insulation efficiency of a wall depends primarily on the surface density (weight per surface unit), the separation of each barrier element, the use of resilient mounting material, the addition of some sound absorbing material between the barrier elements and the elimination of any noise transmission paths. Concrete and masonry walls, because of their relatively high surface density, are excellent construction materials for sound insulation. The acoustical performance of these walls follows closely the prediction by the mass law [3]. The greater is the surface mass, the larger is the sound insulation provided by the partition. For instance, the sound insulation of a masonry concrete wall of 20 kg/m^2 is about 30 dB. A one-brick wall of approximately 23 cm thick has a mass of 415 kg/m^2 and gives a sound insulation of about 50 dB [6].

It should be noticed that a double-leaf construction offers a higher sound insulation than a single-leaf construction with the same total mass. The sound insulation of a double-leaf partition depends on the physical properties of each of the leaves and on the nature of the connections between them. The greater the separation and the less the linkage between the two leaves, the better will be the sound insulation. The transmission of sound via the structural framing can be reduced if a resilient mounting system is used for at least one of the leaves. The inclusion of a sound absorbing material (fiberglass) within the cavity of a double-leaf construction can improve the sound insulation, but the improvement is not usually significant if the two leaves have rigid connections [15].

6.2 Window acoustical insulation

To achieve high sound insulation a structure should not include any light-weight, openable elements such as doors and windows as their generally poor insulation will limit the acoustic performance of the façade. However, buildings are rarely designed in this manner as windows are considered to be important from the practical and aesthetical point of view, allowing for natural lighting, ventilation and visual contact with the external environment [15].

Figure 7 shows the sound insulation values for three different thicknesses of single glass. Increasing the thickness of the glass improves the performance in the low frequency range but it also leads to the dip occurring at an even lower frequency (caused by the coincidence effect explained above). Thus the reduction of the traffic noise over the whole frequency range is not greatly improved when only the thickness of the glass is increased. The coincidence effect is not as great in the performance of laminated glass because the presence of the interlayer affects the bending wave in the glass panel.

A double-leaf construction can provide significant improvements in sound insulation. An important factor determining the effectiveness of a double glazing system is the spacing between the component panels. In order to improve the isolation the best solution is to use double window systems, where the panels are mounted in separate frames. Installation of absorbent material around the frame reduces the effects of resonances in the cavity between the two panels of glass and so further improves the sound insulation. Obviously, the sound insulation decreases when a window is opened for ventilation. The decrease for a staggered opening arrangement (indirect air pathway) is much lower than for an ordinary sliding window, although the ventilation rate is also reduced. Tightly closed or sealed windows can no longer be used for natural ventilation. In this case, a mechanical ventilation or air conditioning system must be provided. Such systems should be carefully chosen so that they produce adequate ventilation without unacceptable noise. The associated air vents should be located away from the roadside façade or should have baffles so that they do not provide paths for the transmission of sound.

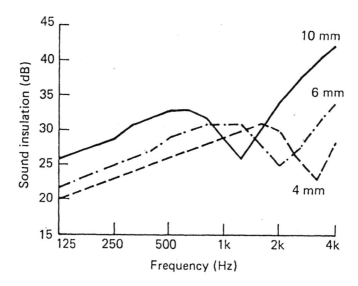

Fig. 7: Sound insulation values for three different thicknesses of glass [15].

6.3 Door acoustical insulation

Doors generally provide low sound insulation (generally lower than walls) so it is preferable if they are located on a building façade not directly exposed to the source of external noise (road traffic, railway, etc.). If a door must be located on the exposed façade its area should be kept as small as possible and it should be fitted with a gasket which provides a good seal all around the door. In particular, the use of undercut doors to accommodate carpets or to allow return air movement can destroy an effective acoustical insulation design. Ideally, the doors should have no openings in them (such as letter slots) and they should not lead directly into a room which is sensitive to noise.

With respect to construction details, it should be noticed that solid doors are generally better sound barriers than hollow doors. Typical values for the sound insulation of doors range from 10-15 dBA (hollow doors) to 15-20 dBA (solid doors). A practical method of raising the acoustical performance of a door is to include one or more layers of sheet lead on the inner surfaces. In addition, filling the cavity with some absorbing material also improves the insulation performance. Where a high level of acoustical insulation is required, double doors back to back with gaskets or the construction of a sound lock should be considered. It should be emphasized that sliding doors rarely provide effective sound insulation due to the difficulty in maintaining a tight acoustical seal at the jamb, head and sill [58].

6.4 Floor acoustical insulation

Impact sound is usually defined as a special kind of structure-borne sound. Perhaps the most common source is footsteps, but there are many others, such as furniture being moved or cleaning equipment operating directly on the surface of floors. In multi-storied buildings, impact sound is radiated into rooms below, but besides this most important case, horizontal transmission between different rooms or dwellings, or even transmission to the rooms above, can also cause serious annoyance to the residents.

The impact sound insulation of concrete floors is never satisfactory unless improved by a soft floor covering or a floating floor system. Suspended ceilings generally reduce impact and airborne sound transmission to a certain extent, depending on the elasticity of the hangers. An obvious solution to the problem of impact sound is to reduce the excitation of the structural floor by covering it with a resilient layer. It should be noticed that these layers do not significantly improve airborne sound insulation; any improvements are usually only at high frequencies. Flexible layers of an uniform material such as plastic or rubber provide only a small sound insulation improvement of about 5-10 dB. By using soft floors composed of a hard upper layer (plastic, rubber or linoleum) and a resilient lower layer (fibre board or foamed plastic) higher impact sound insulations can be obtained, about 15-20 dB. The combination of a soft underpad and a carpet can provide an increase of around 40 dB or more. When a soft floor covering may not be used, the most practical means to increase impact sound insulation is to use a floating floor construction. This floor offers the additional advantage that it will also improve airborne sound insulation [58].

7 Noise control regulations

The policies developed to fight against environmental urban noise include a wide series of administrative measures such as standards for individual sources emission, regulations for immission of noise in buildings based on noise quality criteria, regulations on land use planning, legislation on economic parameters, normatives on operational procedures of the various transportation vehicles, and legislation on research and development.

Noise control regulations can be promoted at an international, national, regional or local level, depending mainly on the specific circumstances existing in each situation (political, economic, social, etc.) and the attitude of the respective bodies towards the environmental noise problems. Actually, the problem is not the shortage of good regulations but slack enforcement. Noise regulations not only are poorly coordinated, but they are often not properly implemented. Quite frequently, the available resources for noise abatement programs are insufficient, so that any large-scale action becomes a very complex task. Another most intractable factor lies in the problem that national governments often pass on the responsibility for enforcing regulations to local authorities, which, in many cases, lack the necessary personal and technical means to undertake an adequate noise control policy [5][35].

Obviously, it is not possible to present here even a superficial account of the endless number of laws, standards and regulations published at different administrative levels around the world, especially considering that the philosophy and contents of existing regulations show many differences from one country to another [61]. This subject will be illustrated only through a short review of some significant topics.

7.1 Noise regulations in Europe

To a greater or lesser extent, the various European countries have introduced different regulations, guidelines or recommendations in order to protect their population against immission from industrial and commercial noise. The rating principle of ISO 1996/2-1987 is applied in nearly all countries. According to this standard, rating levels are calculated from the equivalent continuous sound level Leq plus adjustments for tonal components and impulsive noise. Some countries use additional adjustments, for instance, for immissions during rest periods or for special noise sources. Despite the general agreements in the basic rating principle, there are large differences in detail in the procedures. These differences will make difficult a harmonization of the existing noise regulations on the European level [5][62].

Over the past three decades, the environmental noise control policy developed in Europe (at the European Union level) has essentially consisted of legislation fixing the maximum sound levels for vehicles and machines. This legislation has been basically conceived with single market purposes rather than as an instrument for a general noise abatement program. The corresponding Member States have also enacted a multitude of supplementary regulations and other measures aiming to reduce the impact of noise problems [41].

The original European Community legislation governing sound levels emission for motor vehicles (cars, lorries and buses) was first adopted in 1970 (Directive 70/157/EEC) and has since been amended nine times. The latest amendment by Directive 92/97/EEC came into force in 1996. The noise emission limits established for all motor vehicles have been considerably reduced over these years. So, for passenger cars, such limits were 83 dBA (1972), 80 dBA (1982), 77 dBA (1988/90) and 74 dBA (1995/96). For urban buses, the limits were 89 dBA (1972), 82 dBA (1982), 80 dBA (1988/90) and 78 dBA (1995/96). For heavy lorries, the limits were 91 dBA (1972), 88 dBA (1982), 84 dBA (1988/90) and 80 dBA (1995/96). As the noise limits have fallen, tire/road noise has become more significant and with the new limits this will be the main noise source at speeds above 50 km/hr. Consequently, it has been pointed out that without adequate action addressed to reducing tire/road noise, a further lowering of the present noise limits would not be effective [5].

The European legislation setting limits for the permissible sound level of motorcycles has been in existence since 1978 (Directive 78/1015/EEC). Since then, this legislation has been amended on several occasions in order to introduce lower limit values. For instance, for a motorcycle with less than 80 cm^3, the emission limits were 78 dBA (1980), 77 dBA (1989) and 75 dBA (1996). For a motorcycle with >175 cm^3, these limits were 83-86 dBA (1980), 82 dBA (1989) and 80 dBA (1996). The above noise limits are quite similar to those existing in other many countries around the world.

The United States Government has regulated the noise produced by medium and heavy duty trucks, and also motorcycles and mopeds. Passenger cars, buses and other light vehicles are regulated by certain states and local jurisdictions [63]. To mention only one example, Chicago's ordinance establishes that, for new vehicles operating over 35 mph (56 km/h), the limits are 75 dB for cars, 78 dB for motorcycles, 84 dB for trucks and 77 dB for buses [64].

Regarding air transport European Community regulations, it should be mentioned that the Directive 92/14/EEC, which came into force in April 1995, is the latest in a series of legislative measures beginning in 1979 (Directives 80/51/EEC and 89/629/EEC) aimed at limiting the impact of aircraft noise. Like similar legislation existing in many other states (most of the non-EU European countries, Japan, Australia, New Zealand and the USA), these Directives use the standards specified by the International Civil Aviation Organisation (ICAO). The limit values for individual aircraft types during the take-off and landing operations are usually specified in terms of Effective Perceived Noise Levels (EPNL), and depend on the aircraft weight and number of engines. According to this legislation, the oldest and noisiest jet transport aircraft models (first generation) have been excluded from airports for several years. Heavy aircraft over 25 years old have been banned (with some exemptions) from European Community airports since April 1995. As of April 2002 only aircraft of the third generation (much more quieter) will be allowed to use Community airports. These measures will produce a substantial reduction in the number of people exposed to aircraft noise exceeding 55 dBA [5].

The European Community has also devoted much attention towards the control of noise produced by different types of equipment generally used in construction and public works. The policy consisted in a number of Directives relating to permissible noise emission values, noise test codes and the labeling of equipment with its guaranteed noise emission values (see Section 5.2).

However, the existing Directives only refer to a very small range of noisy outdoor equipment and in recent years there have been many calls from several Member States to extend the present legislation to cover other products and to ensure that national legislation on this subject do not lead to restrictions on trade and cause problems for the functioning of the single market. In February 1998, the European Commission adopted a proposal for a European Parliament and Council Directive on the approximation of the laws of the member states related to the noise emission by equipment used outdoors. It intends to simplify legislation, to contribute to the smooth functioning of the internal market, and to protect the health and well-being of citizens by reducing the overall noise exposure. The proposal covers 55 types of equipment (construction machinery, gardening equipment, special vehicles, etc.), all of which are to be marked with the guaranteed noise emission level [65].

7.2 Noise Quality Criteria

A most important point regarding the noise control regulations refers to the denominated Noise Quality Criteria [66]. These criteria establish the noise levels that are considered as acceptable, for instance, for speech communication, for different uses of buildings, for sleep or for different land uses. Only a few examples of these criteria will be mentioned in this Section.

In particular, the US Department of Housing and Urban Development (HUD) has established regulations for noise abatement and control affecting the approval of new or rehabilitated buildings constructed with financial assistance from the US Government. The degree of acceptability of the noise environment at the proposed site is determined from outdoor, day-night, average A-weighted sound level Ldn. The upper threshold of the "normally acceptable" Ldn noise level is 65 dBA provided there are no loud impulsive noises. If Ldn exceeds 65 dBA, the site is considered as "normally unacceptable". If the noise level is between 65 and 70 dBA, 5 dBA of sound attenuation in addition to that of normal construction must be provided, and 10 dBA additionally for noise levels from 70 to 75 dBA. Finally, for values above 75 dBA, the site is considered as "absolutely unacceptable" [67].

The US Environmental Protection Agency (EPA), concerned only with protection of public health and welfare over life spans, has identified limited Ldn levels in both indoor and outdoor spaces. For instance, the Ldn levels are 45 dBA for indoor activity interference and 55 dBA for outdoor activity interference (residences, hospitals and educational buildings). It should be observed that although the EPA's values are about 10 dBA lower than HUD's normally acceptable levels, they are several decibels too high for good speech intelligibility [68]. In a similar sense, the World Health Organization (WHO) consider Leq values lower than 50/55 dBA (day) and 45 dBA (night) as acceptable noise levels in outdoor environments. Obviously, the existing situation in most urban zones is quite far from all these values, which should be considered as long-term goals, not as standards.

In any case, these specific guidelines are being increasingly integrated into national noise abatement laws. In particular, many European countries have published regulations or recommendations aiming for immission limits in noise sensitive areas closely similar to the above values. This is the case, for instance, with the Spanish Basic Normative of Construction, first published in 1981 and last revised in 1988. Regarding residential buildings, the values of equivalent sound levels Leq measured in living rooms should not exceed 45 dBA during daytime (from 08.00 to 22.00 hours) or 40 dBA during nighttime (from 22.00 to 08.00 hours). The corresponding Leq values for bedrooms are 40 and 30 dBA, respectively [69].

8 Economic instruments

Legislation to control environmental noise has conducted, in many countries, to the development of a wide variety of regulations and standards. The contents of these guidelines involve many social and economic factors and the administration that promotes them (at one or another level) has to be aware of all these consequences. This Section examines the various economic instruments which can be used for noise abatement. In that sense, monetary noise charges are the possibility most widely cited, but other initiatives such as noise damage compensation or promotion of quieter products have been also implemented [70].

8.1 Noise charges

In purely economic terms, environmental noise produces always an uncompensated social cost. It has been suggested that if a noise-producing activity had a price which would have to be paid by the agent responsible for the activity, market decisions would reflect the cost of the noise. Giving a price to noise would induce noise producers either to reduce noise or to pay for the noise they produce through monetary charges [71].

In general, noise charges can be defined as a payment to the appropriate authority for each unit of noise exceeding a certain level emitted into the environment (level established by a prevailing legal regulation). Consequently, monetary charges put a price on noise. The effectiveness of this instrument in achieving noise abatement depends on whether the relevant markets operate in such a way that the response of noise makers is as required [72].

The basic objective of charges is to induce noise makers to curtail noise in order to reduce the cost of fines. In addition, the funds collected in this way may be redistributed for noise abatement purposes. The redistributive function can take various forms, such as the total or partial financing of collective noise control facilities, the contribution to noise control investments made by the noise maker, or the payment of some monetary compensation to the affected people. The noise charges initiatives can also act as an important stimulus to technical progress, helping to develop more and more quieter products.

Monetary penalties are not a recent approach to pollution control. Actually, this strategy has been in existence for a long time for the control of water and solid waste pollution in many countries In our case, to be effective, fines should be related as closely as possible to the noise emission and impact created on a given community. Experience has shown that monetary charges and direct noise controls are not incompatible, but they must be used in combination as complementary noise abatement instruments.

It is sometimes argued that a monetary penalty is only a way of paying for the right to pollute and that excessive noise should simply be prohibited instead of compensated. This argument has force as long as the charge is set at a too low level, so it is cheaper for the noise maker to pay the fine than to reduce the noise. Obviously, if the noise charges are set at a level that is high enough to act as an incentive for the reduction of noise, the problem disappears.

Another major argument against monetary charges is that their effectiveness in reducing noise levels remains uncertain, whereas adequate regulation fixes a compulsory and precise limit level. Regarding this argument, it should be borne in mind that some of the uncertainty associated with charges could be removed if they are set high enough to ensure the fulfillment of the proposed objectives. Furthermore, monetary penalties can be combined with regulations in such a way that fines provide financial inputs for the achievement of the objectives set by the regulations [72].

A field in which these charges have been widely applied in recent years is aircraft noise. The basic philosophy of this strategy is that the aircraft operators pay a fee proportionate to the noise that they generate. The noise charge is applied by the airport authorities as an additional landing fee. In that sense, the charge acts as a complement to existing international standards or local anti-noise procedures (landing and take-off procedures, restrictions on airport use, etc.) established to limit aircraft noise emissions. The aircraft noise charges have three main functions: to encourage airlines to replace their noisiest models (or assign these models to operations serving locations where such charges do not exist), to encourage aircraft manufacturers to produce quieter aircraft, and to contribute to the financing of soundproofing of buildings, rehousing residents or any other measures aimed at protecting airport neighbors from excessive noise [73].

European airports have shown leadership in establishing noise-based charges as valid instruments to control aircraft noise. The use of noise fines was first introduced in Europe in the 1970s and have been growing since then. In a recent survey addressed to 99 European airports, 29 of them reported having had noise related charges and some 27 more indicated that such charges were planned in the near future as an instrument to influence the use of quieter aircraft [5]. In France, the scheme initially applied to Orly and Charles de Gaulle airports (Paris) was extended in 1983 to all airports. In this case, the landing fee applied to all aircraft was reduced or increased according to the respective noise level emission. This scheme created funds for some noise abatement measures such as improving the acoustic insulation of buildings, or generated a plan for the acquisition and demolition of dwellings located in the most exposed areas. Some other countries, such as the United Kingdom or Germany, apply a certain reduction of the landing fee at commercial airports for aircraft complying with the International Civil Aviation Organisation (ICAO) standards. Obviously, as more and more airports apply some type of budgetary charges to limit aircraft noise emissions, the greater will be the incentive on companies and manufacturers to satisfy these demands.

The application of a similar scheme of noise fines to road vehicles raises some problems. Firstly, who will provide the payments, the manufacturer or the owner of the vehicle? In the first case, the initiative would influence the design of quieter vehicles and the application of the charge would be easier in practice (clearly identified and controlled transactors). The main disadvantage of this approach is that vehicle noise potential impact not only depends on the characteristics of the vehicle (power) but also on the driving style adopted by the user. In other words, a quiet vehicle can cause considerable noise impact if driven noisily along a quiet urban area at night. However, the implementation of a proper noise charge to each individual driver and vehicle is an impossible proposition.

The application of a vehicle noise tax levied annually (for example) could help to ensure that vehicles are properly maintained in use. This alternative could be combined with a sliding scale applied to new vehicles, penalizing the noisier ones. This scheme would benefit both the manufacturer and the owner of road vehicles. The owner would prefer to use quieter vehicles because the corresponding taxes would be lower and the manufacturer would be encouraged to produce quieter vehicles because this policy would increase sales.

Several methods of assessing the tax charges have been considered. A specially simple method could be based on defining for each category of vehicles a threshold value below which vehicles would not be taxed, while each decibel in excess of such a threshold value would be charged at some established rate (lineal or exponential). The establishment of a noise charge in the form of a taxation on vehicle fuel, in force in the Netherlands since 1980 to help finance the noise abatement program in that country, represents a different approach to this problem. Obviously, this tax is based on the actual use of the vehicle and it is not really linked to its intrinsic noise output.

In any case, the present experience with charging schemes on road vehicle noise is much more limited than on aircraft. Practical difficulties in the design and implementation of such schemes, as well as political difficulties, have so far prevented its application [72].

8.2 Other economic incentives

As a complement to regulations and noise charges, some countries have adopted a wide variety of financial incentives, such as promotion of low-noise products (road vehicles, industrial equipment, building machinery, home appliances, etc.), financial support to research related with noise control (manufacturers, consulting firms, university laboratories, etc.), or allowance of some monetary compensation to those suffering from noise [72].

Restrictions on the use of especially noisy vehicles at certain times and at certain locations have been put into operation in different countries. However, when exemptions from restrictions are granted for very quiet vehicles, such a regulation acts as an economic incentive because it promotes the production and purchase of quiet vehicles.

Another type of budgetary incentive is the financial assistance granted by public administrations to the research organizations developing low-noise equipment. Some years ago, the Environmental Agency of Germany sponsored a wide research program for producing industrial prototype vehicles with low-noise emission. This program conducted the development of some very quiet vehicles (a car of 73 dBA and a heavy goods vehicle of 77 dBA). An alternative possibility consists in the granting of subsidies stimulating the purchase of quiet vehicles (especially buses and lorries). This strategy has been applied in the Netherlands since 1980 [8].

Noise compensation is defined as a payment (in cash, goods or services) designed to indemnify a person for the damage produced by noise. In a wide sense, noise compensation can be considered as an economic incentive: if the noise producers (highway administration, airports authority, etc.) know that they may be enforced to pay compensation, they will act in such a way as to minimize this payment, either by not producing a noise which may generate a request for compensation from the affected people or by including the costs of such compensation in their investment calculations.

The need for noise compensation can be particularly important in existing built-up areas around airports and along highways. In situations where no satisfactory solution can be adopted, the only possibility may be to provide some type of compensation to affected people. Obviously, priority should be always given to prevention measures. Compensation should be considered as a last resort when all other possibilities have been tried.

Some countries (France, Germany, the United Kingdom, the Netherlands, Japan, etc.) have adopted special provisions in their legislation to allow the granting of such compensation (usually in the form of sound insulation of the affected properties). For example, from 1974 to 1979, the city of Munich (Germany) spent about 20 million marks on insulating buildings from traffic noise (this cost represents only 50% of the actual cost, the other 50% was shared by the dwelling owners). Since 1972 a system introduced for Heathrow Airport (United Kingdom) established the payment of 100% of insulation costs for dwellings situated in the zone where Leq exceeds 77 dBA. The aircraft noise charges collected in the Amsterdam Airport (the Netherlands) were used to finance the soundproofing of the most exposed dwellings [72].

References

[1] Wilson Committee, *Noise - Final Report*, Her Majesty Stationery Office, London, 1963.

[2] OECD, *Fighting Noise in the 1990's*, Organisation for Economic Cooperation and Development, Paris, 1991.

[3] Harris, C.M. (ed.), *Handbook of Acoustical Measurements and Noise Control*, McGraw-Hill, New York, 1991.

[4] Beranek, L.L. & Ver, I.L., *Noise and Vibration Control Engineering*, John Wiley and Sons, New York, 1992.

[5] European Commission Green Paper, *Future Noise Policy, Noise/News International*, vol. 5, no. 2, pp. 77-98, 1997.

[6] Smith, B.J., Peters, R.J. & Owen, S., *Acoustics and Noise Control*, Longman, London, 1982.

[7] Nelson, P.M., Introduction to Transport Noise, Chapter 1, *Transportation Noise. Reference Book*, Butterworths, London, pp. 1/1-1/14, 1987.

[8] Favre, B.M. & Tyler, J., Quiet Vehicle Development, Chapter 8, *Transportation Noise. Reference Book*, Butterworths, London, pp. 8/1-8/16, 1987.

[9] Tyler, J.W., Sources of Vehicle Noise, Chapter 7, *Transportation Noise. Reference Book*, Butterworths, London, pp. 7/1-7/39, 1987.

[10] Crocker, M.J., Generation of noise in machinery, its control, and the identification of noise sources, Chapter 66, *Handbook of Acoustics*, Wiley and Sons, New York, pp. 815-848, 1998.

[11] Federal Environmental Agency, *Noise Abatement. Major Research Topics*, Federal Minister of the Interior, Bonn, 1985.

[12] Sandberg, U., Vehicle noise emission changes during the period 1974-1988, *Proc. International Conference on Noise Control Engineering (Internoise'89)*, Institute of Noise Control Engineering, Washington, pp. 711-714, 1989.

[13] Verdan, G., Public aspects of noise control, *Proc. 6th International FASE Congress*, Swiss Acoustical Society, Zurich, pp. 249-258, 1992.

[14] Favre, B., Factors Affecting Traffic Noise and Methods of Prediction, Chapter 10, *Transportation Noise. Reference Book*, ed. P.M. Nelson, Butterworths, London, pp. 10/1-10/24, 1987.

[15] Buna, B. & Burguess, M., Methods of Controlling Traffic Noise Impact, Chapter 11, *Transportation Noise. Reference Book*, ed. P.M. Nelson, Butterworths, London, pp. 11/1-11/26, 1987.

[16] Alexandre, A., European efforts to reduce the impact of traffic noise, *Proc. International Conference on Noise Control Engineering (Internoise'72)*, Institute of Noise Control Engineering, Washington, pp. 208-211, 1972.

[17] Buna, B., Some Characteristics of Noise From Single Vehicles, Chapter 6, *Transportation Noise. Reference Book*, Butterworths, London, pp. 6/1-6/13, 1987.

[18] Abbot, P.G. & Layfield, R.E., The change in traffic noise levels following the installation of speed control cushions and road humps, *Proc.*

International Conference on Noise Control Engineering (Internoise'96), Institute of Acoustics, St. Albans, pp. 1957-1962, 1996.

[19] Migneron, J.G., *Acoustique urbaine*, Masson, Paris, 1980.

[20] Woehner, H., Sound propagation at tunnel openings, *Noise Control Eng. Journal*, vol. 39, pp. 47-56, 1992.

[21] Williams, R., Bachman, T., Blockland, G., Fingerhut, H.P., Hamet, J.F., Sandberg, U. & Taylor, N., Bruit de contact pneu-chausée: Etat de l'art, *Acoustique et Techniques*, no. 15, pp. 17-31, 1998.

[22] Sandberg, U., Low noise road surfaces, *Proc. Conference on Noise in Metropolitan Cities*, Spanish Acoustical Society, Madrid, pp. 131-167, 1991.

[23] Sandberg, U., A new porous pavement with extended acoustical lifetime and useful even on low-speed roads, *Proc. International Conference on Noise Control Engineering (Internoise'97)*, OPAKFI, Budapest, pp. 99-104, 1997.

[24] Meier, A.V., Acoustically optimized porous road surfaces. Recent experiences and new developments, *Proc. International Conference on Noise Control Engineering (Internoise'88)*, Société Française d'Acoustique, Paris, pp. 1323-1326, 1988.

[25] Sandberg, U. & Ejsmont, J.A., Texturing of cement concrete pavements to reduce traffic noise, *Noise Control Eng. Journal*, vol. 46, pp. 231-243, 1998.

[26] Tai, D.M.K. & Szeto, W.K., Hong Kong's experience in tackling traffic noise from new roads. Direct and indirect mitigation measures. *Proc. International Conference on Noise Control Engineering (Internoise'97)*, OPAKFI, Budapest, pp. 847-850, 1997.

[27] García, A., Garrigues, J.V. & Romero, J., Resultados de un nuevo estudio de las condiciones acústicas del Campus Universitario de Burjassot (Valencia), *Proc. Jornadas Nacionales de Acústica (Tecniacústica'93)*, Spanish Acoustical Society, Madrid, pp. 7-10, 1993.

[28] Leroux, M., Les écrans acoustiques: perception et représentations des riverains, *Acoustique et Techniques*, no. 16, pp. 29-32, 1998.

[29] Maekawa, Z., Noise reduction by screens, *Applied Acoustics*, vol. 1, pp. 157-173, 1968.

[30] Maekawa, Z., Acoustic Shielding: Noise Reduction by Thin and Wide Barriers, Chapter 3, *Noise Pollution*, John Wiley and Sons, New York, pp. 133-145, 1986.

[31] Fujiwara, K. & Furuta, N., Sound shielding efficiency of a barrier with a cylinder at the edge, *Noise Control Eng. Journal*, vol. 37, pp. 5-11, 1991.

[32] Harris, R.A., Vegetative barriers: an alternative highway noise abatement measure, *Noise Control Eng. Journal*, vol. 27, pp. 4-8, 1986.

[33] Smith, M.T.J. & Williams, J., Subsonic Aircraft Noise, Chapter 18, *Transportation Noise. Reference Book*, Butterworths, London, pp. 18/1-18/36, 1987.

[34] Bragdon, C.R., Control of Airport Noise Impact, Chapter 20, *Transportation Noise. Reference Book*, ed. P.M. Nelson, Butterworths,

London, pp. 20/1-20/21, 1987.

[35] Miller, R.L., Federal regulations and other activities in noise control, *Noise Control Eng. Journal*, vol. 44, pp. 149-152, 1996.

[36] Coleman, L. & Eldred, K.M., Fifteen years of noise control at Logan International Airport, *Proc. International Conference on Noise Control Engineering (Internoise'89)*, Institute of Noise Control Engineering, Washington, pp. 655-658, 1989.

[37] The Noise Advisory Council, *Aircraft Noise: Flight Routing Near Airports*, Her Majesty's Stationery Office, London, 1971.

[38] Nordic Noise Group, *Reduction of aircraft noise at Nordic airports before the year 2000*, Nordic Council of Ministers, Copenhagen, 1987.

[39] Eldred, K.M., Airport noise, Chapter 70, *Handbook of Acoustics*, John Wiley and Sons, New York, pp. 883-896, 1998.

[40] Herzing, W., Noise abatement considerations while planning the new Munich Airport, *Proc. International Conference on Noise Control Engineering (Internoise'86)*, Institute of Noise Control Engineering, Washington, pp. 935-940, 1986.

[41] Wolfe, S.L., Introduction to Train Noise, Chapter 13, *Transportation Noise. Reference Book*, Butterworths, London, pp. 13/1-13/6, 1987.

[42] Remington, P.J., Kurzweil, L.G. & Towers, D.A., Practical Examples of Train Noise and Vibration Control, Chapter 17, *Transportation Noise. Reference Book*, Butterworths, London, pp. 17/1-17/23, 1987.

[43] Stanworth, C., Sources of Railway Noise, Chapter 14, *Transportation Noise. Reference Book*, Butterworths, London, pp. 14/1-14/13, 1987.

[44] Thompson, D.J., Jones, C.J.C. & de France, G., The effects of the rail support stiffness on railway rolling noise, *Proc. Forum Acusticum 1999*, Berlin, *Acustica / Acta Acustica*, vol. 85, Suppl. 1, S21, 1999.

[45] Remington, P.L., The resiliently treated wheel - A concept for control of wheel/rail rolling noise, *Proc. International Conference on Noise Control Engineering (Internoise'88)*, Société Française d'Acoustique, Paris, pp. 1401-1404, 1988.

[45] Dings, P., Measures for noise reduction on steel railway bridges, *Proc. International Conference on Noise Control Engineering (Internoise'97)*, Institute of Noise Control Engineering, Washington, pp. 143-146, 1997.

[47] Jonasson, H.G., Sound reduction of low railway barriers, *Proc. International Conference on Noise Control Engineering (Internoise'97)*, OPAKFI, Budapest, pp. 417-420, 1997.

[48] Tyler, J.W., Noise from heavy and special vehicles, *Proc. Conference on Noise in Metropolitan Cities*, Spanish Acoustical Society, Madrid, pp. 3-28, 1991.

[49] Braunschweig, G.V., Noise of public works, *Proc. Conference on Noise in Metropolitan Cities*, Spanish Acoustical Society, Madrid, pp. 171-178, 1991.

[50] Asselineau, M., Noise reduction of a building yard in an urban area: a case study, *Proc. International Conference on Noise Control Engineering (Internoise'97)*, OPAKFI, Budapest, pp. 943-946, 1997.

[51] Axelsson, A., Recreational exposure to noise and its effects, *Noise Control Eng. Journal*, vol. 44, pp. 127-134, 1996.

[52] The Noise Advisory Council, *Noise in Public Places*, Report by a Working Group of the Council, Her Majesty's Stationery Office, London, 1974.

[53] Granneman, J.H. & Huizer, H., Noise emission from stadiums, *Proc. International Conference on Noise Control Engineering (Internoise'96)*, Liverpool, pp. 1879-1882, 1996.

[54] Griffits, J.E.T., Hinton, J. & Morris, P., Noise Council Code of Practice on Environmental Noise Control at Concerts. Case studies, *Proc. International Conference on Noise Control Engineering (Internoise'96)*, Institute of Acoustics, St. Albans, pp. 1887-1892, 1996.

[55] Boxall, J.E., Tromp, F., Ho, S.W.F., Chan, K.S. & Ng, P.S., Outdoor entertainment in Hong Kong. A Noise Balancing Act, *Proceedings International Conference on Noise Control Engineering (Internoise'96)*, Institute of Acoustics, St. Albans, pp. 1919-1892, 1996.

[56] Lemasle, L., Les nuisances sonores dues à l'exploitation des bars et des discothèques, *Acoustique et Techniques*, no. 16, pp. 33-36, 1998.

[57] García, A., Albero, V., Calvo, F., Romero, J., Marcos, A. & Sánchez, J., Estudio del impacto acústico producido por los pubs y discotecas de la Comunidad Valenciana, *Revista de Acústica*, vol. 26, pp. 5-12, 1996.

[58] Warnock, A.C.C. & Fasold, W., Sound insulation: airborne and impact, Chapter, 76, *Handbook of Acoustics*, John Wiley and Sons, New York, pp. 956-972, 1998.

[59] Wetherill, E.A., Control of noise and vibration in dwellings: a practical alternative, *Proceedings International Conference on Noise Control Engineering (Internoise'87)*, Acoustical Society of China, Beijing, pp. 695-698, 1987.

[60] Kihlman, T., Fifty years of "development" in sound insulation of dwellings, *Noise Control Eng. Journal*, vol. 42, pp. 47-52, 1994.

[61] Aecherli, W., Judicial and Legal Aspects of Noise Control, Chapter 18, *Noise Pollution*, John Wiley and Sons, New York, pp. 419-429, 1986.

[62] Gottlob, D.P.A., Regulations for industrial noise immissions in European countries, *Proc. Forum Acusticum 1996*, Antwerp, *Acustica / Acta Acustica*, vol. 82, Suppl. 1., S98, 1996.

[63] Hickling, R., Surface transportation noise, Chapter 71, *Handbook of Acoustics*, Wiley and Sons, New York, pp. 897-905, 1998.

[64] Chicago Municipal Code, Article 4, *Noise and Vibration Control*, Paragraphs 17-4.1 to 17-4.30, 1984.

[65] Irmer, V.K.P., A new EU Directive concerning the noise emission by equipment used outdoors, *Proc. Forum Acusticum 1999*, Berlin, *Acustica / Acta Acustica*, vol. 85, Suppl. 1., S451, 1999.

[66] Beranek, L.L., Applications of NBC and RC noise criterion curves for specification and evaluation of noise in buildings, *Noise Control Eng. Journal*, vol. 45, pp. 209-216, 1997.

[67] Galloway, W.J. & Schultz, T.J., *Noise Assessment Guidelines*, Report HUD-CPD-586, US Department of Housing and Urban Development,

Washington, 1979.

[68] Environmental Protection Agency, *Information on Levels of Environmental Noise Requisites to Protect Health and Welfare with an Adequate Margin of Safety*, Report No. 550/9-74-004, US Environmental Protection Agency, Washington, 1974.

[69] Ministerio de Obras Públicas y Transporte, *Norma Básica de la Edificación, NBE-CA-88, Condiciones acústicas en los edificios*, MOPT, Madrid, 1988.

[70] Barde, J.P., Noise abatement policies in OECD countries: an assessment, *Proc. Conference on Noise in Metropolitan Cities,* Spanish Acoustical Society, Madrid, pp. 189-200, 1991.

[71] Alexandre, A. & Barde, J.P., The Evaluation of Noise, Chapter 23, *Transportation Noise. Reference Book*, Butterworths, London, pp. 23/1-23/10, 1987.

[72] Alexandre, A. & Barde, J.P., Economic Instruments for Transport Noise Abatement, Chapter 24, *Transportation Noise. Reference Book*, Butterworths, London, pp. 24/1-24/8, 1987.

[73] Burguess, M., Approaches to aircraft noise amelioration schemes, *Proc. 14th International Congress on Acoustics*, Acoustical Society of China, Beijing, vol. 2, E2-9, 1992.

WIT_PRESS_

Boundary Element Acoustics

Fundamentals and Computer Codes

*Editor: **T.W. WU**, University of Kentucky, USA*

Using this unique tutorial readers will not only become familiar with the basic principles of the BEM in acoustics, but will also be able to gain hands-on experience by constructing computer codes for a wide range of problems in the field.

The first text to cover virtually every aspect of the BEM in acoustics, this book will be ideally suited to researchers, engineers and graduate students. A CD-ROM providing FORTRAN codes is also included.

Contents: Fundamentals of Linear Acoustics; The Helmholtz Integral Equation; Two-Dimensional Problems; Three-Dimensional Problems; The Normal-Derivative Integral Equation; Indirect Variational Boundary Element Method in Acoustics; Acoustic Eigenvalue Analysis by Boundary Element Methods; Time Domain Three-Dimensional Analysis; Extended Kirchhoff Integral Formulations.

Series: Advances in Boundary Elements
ISBN: 1-85312-570-9   2000
256pp + CD-ROM   £95.00/US$149.00

Lecturers
Price reductions are available when you purchase multiple copies of this text. Please contact the Marketing Department at WIT Press for details.

Boundary Elements in Acoustics

*Editor: **O. VON ESTORFF**, Technische Universität Hamburg-Harburg, Germany*

In recent years the application of Boundary Element Methods to acoustical problems has gained much popularity. The methodology is particularly effective and accurate if sound radiation and unbounded acoustic media are involved.

Bringing together chapters from leading university teachers and researchers, as well as researchers in industry, this book provides state-of-the-art reports on all aspects of BEM calculations in acoustics. Special attention is paid to efficiency and accuracy issues, frequency and time domain procedures, direct and indirect formulations, and hybrid as well as inverse techniques. Emphasis is also placed on applications in different fields, such as vehicle acoustics, engine noise, electroacoustic transducers, optimization of musical instruments, environmental protection, and underwater acoustics.

Series: Advances in Boundary Elements
ISBN: 1-85312-556-3   2000
488pp   £159.00/US$246.00

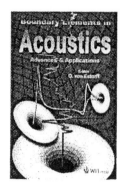

WITPRESS

Active Noise Control

Physical Acoustics and Active Sound and Vibration Control

G. ROSENHOUSE, Technion-Israel Institute of Technology, Haifa, Israel

Active Noise and Vibration Control (ANVC) has to serve disciplines as diverse as civil, mechanical and aeronautical engineering. Lack of understanding and visualization of its consequences may lead to undesired results.

This first volume of **Active Noise Control** pays special attention to this aspect. The author also expands the concept of ANVC in a way that allows for an overwhelming amount of applications within the ANVC frame, and goes far beyond the basic definition of Lueg in 1933. In this way ANVC couples with other areas of physics, such as sonoluminescence, thermodynamics, magnetism and also with areas of biology, such as animal sonars and tinnitus. Such possibilities promote ANVC to unexpected heights and provide fascinating future prospects.

- Features a review of state-of-the-art practice together with more than 1,000 references.
- Summarizes basic linear and nonlinear equations and sound wave analysis and their connections to ANVC.
- Links the material world, including its non-uniformities and boundaries, and ANVC, thus stimulating development of a large amount of topics.

ISBN: 1-85312-373-0; 1-56252-297-3 (US, Canada, Mexico) 2001
424pp £175.00/US$279.00

Computational Acoustics and its Environmental Applications II

Editors: C.A. BREBBIA, Wessex Institute of Technology, UK, J. KENNY, University of Perugia, Italy and R.D. CISKOWSKI, IBM Corporation, USA

Simulation of acoustic behaviour is essential for the design of a wide range of products and living spaces and the prediction of noise in the environment. Computers provide a unique tool for the analysis and design of these problems and have become instrumental in achieving optimum solutions. This book contains the proceedings of the Second International Conference on Computational Acoustics and its Environmental Applications. The papers, which come from leading experts in both academic research and industry, are divided under the following headings: Numerical and Computational Techniques; Aero-Acoustics; Building Acoustics; Wave Propagation; Sound Systems Design and Experiment.

ISBN: 1-85312-459-1 1997
240pp £88.00/US$138.00

WITPRESS

Ecosystems and Sustainable Development

Editors: *C.A. BREBBIA, H. POWER,* *Wessex Institute of Technology, UK and* *J-L. USO, Universitat Jaume I, Spain*

Featuring recent work on the engineering and modelling aspects of ecosystems and sustainable development, this book contains the proceedings of the First International Conference on Ecosystems and Sustainable Development.
Partial Contents: Environmental Policies; Sustainable Development Models; Trade Policy and Development; Natural Resources Management; International Policy Co-ordination; Climate Change, Desertification and Sustainable Development; Climate, Crop and Animal Production.
Series: Advances in Ecological Sciences, Vol 1
ISBN: 1-85312-502-4   1997
704pp   **£195.00/US$295.00**

Ecosystems and Sustainable Development II

Editors: *C.A. BREBBIA, Wessex Institute of Technology, UK and J-L. USO, Universitat Jaume I, Spain*

The tasks facing ecologists now require the collaboration of scientists, engineers, economists and other professionals such as those participating in ECOSUD II, the Second International Conference on Ecosystems and Sustainable Development. Featuring edited versions of the papers presented, this book covers the following topics: Application of Ecological Models in Environmental Management; Biodiversity; Climate Modelling and Ecosystems; Integrated Modelling; Environmental Risk; Sustainable Development Aspects; Lakustrine and Wetlands Ecosystems; Forestation Issues; Computational Modelling of Natural and Human Ecosystems; Natural Resource Management.
Series: Advances in Ecological Sciences, Vol 2
ISBN: 1-85312-687-X   1999
416pp   **£175.00/US$287.00**

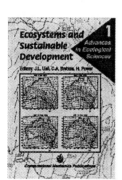

WIT*PRESS*

Risk Analysis II

Editor: C.A. BREBBIA, Wessex Institute of Technology, UK

This book is concerned with many different aspects of computer simulation in risk analysis and hazard mitigation ranging from the specific assessment of risk to mitigation associated with both natural and anthropogenic hazards.

Covering a series of important topics, which are of current research interest and have practical applications, the contributions featured were first presented at the second international conference on this subject. Over 50 papers are included and these are divided under the following headings: Hazard Prevention, Management and Control; Estimation of Risk; Emergency Response; Data Collection and Analysis; Hazardous Materials in Transit; Water Resource Modelling and Management; Landslides; Earthquakes; Soil and Water Contamination; Air Quality Studies; and Case Studies.

Series: Management Information Systems
ISBN: 1-85312-830-9   2000
584pp   £193.00/US$299.00

Computational Acoustics in Architecture

Editor: J.J. SENDRA, University of Sevilla, Spain

Containing a significant amount of state-of-the-art knowledge on room acoustics, this book is written by authors or work teams, all of whom are internationally acknowledged researchers in this field.

The first two chapters centre on the most outstanding aspects of room acoustics studied in depth this century, namely absorption, sound reflection and diffusion, and echo and reverberation. Much current research is dedicated to perfecting models that analyse the so-called subjective attributes of sound fields, and the following sections present studies of simulation models of the binaural experience of listeners in a room. Finally, there are two examinations of recent work carried out on acoustics in concert halls and auditoria, and churches.

Series: Advances in Architecture, Vol 8
ISBN: 1-85312-557-1   2000
192pp   £87.00/US$140.00

CPSIA information can be obtained at www.ICGtesting.com
Printed in the USA
BVOW031654050212

281993BV00005B/19/A